TABULAR STATISTICS

The Statistical Table as a methodological tool

Hugo Casanova

TABULAR STATISTICS

Logic and Dynamics of Statistical Tables

HUGO CASANOVA
TABULAR STATISTICS
Logic and Dynamics of Statistical Tables

estadisticamigable.blogspot.com
casanovade@gmail.com
orcid.org/0000-0001-7393-5163
Cover image. Pixabay.com (Juergen Sieber)
Personal brand logo design by Jorge Casanova

First addition in Spanish, May 2013
Second edition in Spanish, corrected and augmented, January 2024
First English edition, March 2024

Translation aided with Deepl and Google
Mind maps and illustrations with VUE. Visual Understanding environment, free

Dedication

To my parents Herma and Hugo
To my wife Lucía

Acknowledgments

To my hundreds of students at the universities where I have been able to teach and who inspired me to develop the idea of an algorithm for making tables. With them, I practiced the different designs and evaluated their user-friendliness.

To the Escuela Venezolana de Planificación who published the first edition of the book Estadística Tabular. Estructura, tipos y métodos para hacer tablas y gráficos, ISBN: 978-980-7440-71-4. Teaching Notes No 4, May 2013

To the Revista Docencia Universitaria de la UCV, vol. XVIII No 1 and 2 of 2017 for publishing my article complementary to the book SL method of construction of statistical tables as a manual interface of the digital divide in education.

To the Researchgate portal where the article Tabular Logic. Holistic view of structure, analysis and semantic table. CC BY-NC-ND 4.0 license.

To my brother Jorge who took great care in the design of my personal brand.

Table of contents

Foreword

The tables referred to in this book differ from other types of tables, lists or charts. That is, not all content or numerical information that is "framed" in a box limited by lines and formed by cells is a *statistical table*. It differs from the others by a structure that allows it to become a method of analysis and not an information file. The differences have to do with the use of tables as they are technical tools. Lists are very similar to databases and information is obtained from them by means of search and reference commands or logical commands. There are tables that display information such as the results of students' grades, their averages, etc. but they are not statistical tables in a methodological sense, they only display information. Finally there are tabular information that show textual information and that we call *charts*. The statistical table has a structure that not only displays the information, but also shows a certain analysis depending on the arrangement of the variables in the table

The present research was initiated in 2003 on its *own initiative* and *ad hoc*, it was aimed at designing a method of table construction to be explained in class since it was noted the lack of expertise in university students to plan the tabular analysis and then to design the tool with which to perform such analysis. The construction of the table has an important intuitive component because it is related to the heading and the items, but what is difficult is the dynamization of the table when there are several variables, that is, to use it as a method of *Tabular Analysis*.

However, they are already done by statistical programs, many results of operations are provided in tables and the greatest degree achieved has been to dynamize the box in the form of rows and columns without methodological implications. Therefore, the previous phase of the research consisted of a methodological immersion in all these automated tabular processes or exploratory research that concluded in an ad hoc research proposal, namely to associate an external table construction logic understood by researchers, teachers and students so that they could build them with the help of the software and not depend on it, a logic mastered by the researcher. Finally, the method was tested, empirically validated for several years with various groups of students until a first part was published in May 2013 (Casanova, Tabular Statistics. Structure, types and methods for making tables and graphs. Método SL para hacer tablas, 2013), subsequently an update was published (Casanova, SL method of constructing statistical tables as a manual interase of the digital divide in education, 2017) of the approach that is incorporated in this book and finally this more finished proposal.

Caracas, November de 2023

Why a Statistical Tables Book

Two important aspects marked the need to write a *Tabular Statistics*, the first is that the development of statistical software has allowed the dynamization of the statistical table, the second is the inexistence of statistical table books beyond the technical manuals of the software, and both aspects are related. The non-existence of statistical literature on the subject forces the particular use of software manuals that only teach the use of their specific windows. There was no *general method* that encompasses and unifies a tabular method based on a logic that generates a methodology of construction and communication of the process, the programmer communicates with the machine through the software but not with the researcher and this is what this Tabular Statistics intends to do.

Computer development had a double impact on statistics; on the one hand, it systematized many statistical processes that had long and complex iterations and, on the other hand, it forced the systematization and complexification of simple methods. Tables are inserted in this second impact. These were made by simple methods, but the possibilities offered by computer science forced the creation of more complex methods to improve the analysis while creating a difficulty, the design of new statistical methods is made for immediate automation being offered in statistical packages or software. In other words, *technological knowledge* merged with *methodological knowledge*, creating new dependencies.

Some structural and semantic disadvantages of this techno-methodological fusion in statistical tables are, firstly, in some programs their manuals do not define the necessary concepts such as table and subtable, layer, etc., beyond being intuitive elements, known more by culture and statistical tradition, this makes the development of a tabular language and appropriate methodologies impossible. Secondly, the names of the percentages that are calculated do not refer to the predicates, generating confusion; their names are % row, % column, % row layer, etc., very confusing, very confusing; thirdly, only six or eight percentages are obtained referring to totals located in certain places of the table, some justified by tradition, such as those of the borders, and others not, since they could also be calculated on the remaining ones; although these percentages are traditional and useful for tables of three variables, as the size grows by the number of variables, the calculation is maintained on the same totals unjustifiably; on the contrary, if the table is small, the percentages are repeated unnecessarily assuming several names for similar percentages; indicating that they are fixed not dynamic; finally, in fourth place, there is no way to differentiate, nominally, one table from another, neither in its form nor in its content; that is, we could not communicate a tabular requirement; we would not know how to call them or differentiate them, linguistically, from the simplest to the most complex ones.

This generated a *digital gap* in the students that manifests itself in a lack of knowledge of tabular design as a heuristic method and a lack of expertise or skill in the use of statistical technology, causing a negative attitude towards it, believing that it is difficult to use. This lack of knowledge begins with an inadequate conceptualization of tables. On the other hand, the student, when trying to use the statistical program, asks for a basic tabular design of the organization of the variables and, the student, not knowing it, enters in conflict producing a cognitive dissonance, because he confronts two opposing ideas. He has no previous method to make a pre-design of the table and becomes inept in front of the technology because he does not know how to use it. In such a way that developing didactic strategies for the design of tables was necessary to improve the ease of use of the technology and in the end *Tabular Statistics* has already been taught for some years in laboratories and classrooms that achieve *good versatility* by *increasing the ease of use of the information, improving the attitude towards the use of the information* and *promoting the usefulness of the information*, characteristics pointed out as necessary by Young Valera (2004).

Tabular Statistics does not focus on the numbers of the cells but on the semantic ones or items because they are the ones that the researcher seeks to measure, therefore it begins with the redefinition of the frequency advancing towards the combination of items that make the tabular structure more complex and includes a classification of the frequencies that includes the indicators, this is discussed in the first part.

The second part is dedicated to the Tabular Structure, so called because the complexity and dynamics of the tables allowed by computer science require new definitions, the definition of conventional table as a data display acquired a first methodological level with the cross tabulation of contingency tables, however, now the tabular dynamics and the size that tables acquire require adequate definitions so that this part defines the characteristics of a normative table, the typology according to the dimensionality of the structure, some types of tabular structures are evaluated; On the other hand, the naming of percentages also becomes more complex; the traditional names of row, column and total percentages become insufficient in view of the possibilities of nesting and layer formation.

Finally, the third part introduces the necessary elements for the design of a tabular language that allows the teacher and researcher to communicate the tabular requirements and introduces a semantic analysis necessary for the interpretation of the results. This is vital since the diversity of tabular structures having as other objectives the optimization of the size on the sheet requires a logic that allows to know the equivalence between them.

What are Tabular Statistics?

The impact of informatics on Statistics has been definitive, Casanova (2022) explores the relationship between the new Data Intensive Science and Statistics which means the claim of Informatics on the statistical methodology in an unprecedented relaunch, that is the whole statistical methodology now takes on a new context through the Fourt Paradigm or Industry 4.0, see Hey *et all.* (2009), which particularly turns the simple statistical table into a mature method of analysis by the dynamics it imprints on the box or table.

Thus, tabular statistics encompasses the typical *quantitative analysis* of percentages and absolute values with the novel *semantic analysis*[1] that starts from the variables or from the operationalization made by the researcher himself, placing the percentages in function of these, in such a way that it connects the planning of a research with the result. The semantic analysis has several levels by the method of reduction of the intention[2] that allows to relativize the percentages with respect to the evaluated items, also for the dynamics of the box or tabular dynamics it adds a syntactic system that allows a better didactic, that is, it enables teachers to denote a table and differentiate it from others nominally, thus adding a tabular logic through which equivalences are obtained between tables of different sizes

The above implies the development of new definitions of statistical table, holistic table or tabular structure, multiple frequency, incorporates a more complete classification of frequencies because it divides absolute frequencies into direct and derived by arithmetic ratio and relative frequencies not only percentages but from the geometric ratio and proportion, that is, the basis of the indicators, so the indicators in general enter as derived frequencies from the direct ones. On the other hand, the definition of statistical table allows structuring several tables in holistic structures, equivalence between structures, tabular family, tabular class and subclass, so that we are generally in the presence of holistic tabular structures in a tabular dynamic that gives flexibility to the analysis and interpretation of the results.

---

[1] This would be a qualitative analysis based on the predicates or labels of the variables. Generally, the interpretation of the results of quantitative analysis is combined with qualitative analysis. The real qualitative analysis is the semantic analysis of reduction of the intent by extension

[2] We must differentiate intention from intent, the latter is used in logic on the predicates or characteristics of definitions, the former is psychological.

PART ONE

Variables, Frequencies and Quantification

Chapter 1. Variables and Measurement Scales

In this chapter I will present in a practical way the topic of variables and measurement scales. The subject is extremely rich because it includes a large part of the way in which social science has been approaching the approach and solution of many problems. Although it is true that social science goes beyond the simple fact of defining a specific metric, the method of analyzing variables (i.e. separating phenomena into variables) and finding their recurrence is far from being exhausted.

The problem of deciding on metrics, the best scales, how to operationalize the constructs, reliability, etc. are topics of the greatest possible specialization. This text deals only with the basic understanding of variables and scales in the necessary depth. It only seeks to expose them for a minimal but comprehensive learning.

The term "variable" refers to the intuitive idea that something can change some characteristic, not remain fixed or constant

In our context variable will be a set of observations transformed into digits, words, numbers, numerals or codes, by any valid quantitative or qualitative method. However, these observations, insofar as they refer to "something", describe it or say something about it using established concepts. In quantification processes, three basic concepts are used that seek to determine the degree to which we can measure or objectify things by means of instruments designed for that purpose. These instruments range from criteria for classifying to physical instruments that measure the properties (extensive or intensive) of things. **These three concepts are the classificatory, the comparative and the metric.** We will concentrate on the first two, which are of primary use, at least, in modern biosocial science and which are the ones that require the most tabular tools

We warn that not all things or phenomena are possible to treat (quantify) under these three concepts, but it is possible that all things have aspects or characteristics that can be treated with these conceptual tools, for example, neurosis is not something to treat by means of these concepts but it will have aspects that can be treated or conceptualized in this way, it can at least be classified. It is said that all science goes through this primary stage of classifying in order to evolve towards metrization (it is an interesting thing and it is discussed). The subject is mentioned so that the student becomes aware of this).

These three concepts of quantification, classificatory, comparative and metric, are operationalized in three **levels of measurement** or objectification, namely, the nominal, the ordinal and the metric (which is subdivided into two more levels, that of intervals and that of ratio). It is generally believed (countless texts treat it as such) that levels of measurement are one thing and scales[3] are another, an error that is produced by misunderstanding the relationship between ontology and epistemology (between being and knowing). Every scale belongs to a level of measurement

Nominal Level. Qualifying Concepts

If we were to observe the planets Earth, Mars, Neptune, Saturn, etc., as a Whole, each one of them as a totality, without observing any specific attribute (such as distance from the Sun or size of each one) and we wanted to

---

[3] For example, the famous text Kinnear and Taylor (1981), talks about nominal, ordinal and metric scales and then continues with a list of scales that are a subset of the previous ones.

differentiate them, we would simply have to assign them a name or code. The planet we now call Earth, and on which we live, we could call XW25 and it would still be the same planet.

We could call such planets by other names and they would still be the same. That is, Mars can be called "3" or "@". Already in other languages are mentioned differently (but refer to the same planet), Mars (Spanish), المريخ (Arabic), Mart (Catalan), 火星 (traditional Chinese) (translations taken from the Internet). By this we mean that the *name* does not mean anything, it is simply a *code*. The name or code is put only to differentiate the thing from another. Mars is just a name that differentiates that planet from Earth or Neptune. The "name" serves only to classify it. To give a name to something is the most elementary level of classification.

Just as we give proper names to things (as in the case of the planets) we can give names to the attributes of things as general ideas. Thus Red, Hard, etc. are names with which we identify some ideas. Red refers us to an idea of color. Hard also refers us to the idea of that which has that appearance to the touch.

The name refers us to the Whole that we identify with that name, not to any part or degree of it. All reds (more or less red) are Red; Mars is "Mars" not the "4th planet", nor the "rebellious planet", etc.; Pluto is "Pluto" not the "dead planet" nor the "ninth planet", although these senses refer to them as "the morning star" refers to Venus. But they are themselves Mars, Pluto, Red, terms of ideas in a total, complete, unfractionated way.

What differentiates the planets (as a whole, as totalities) is that they are distinct ($\neq$); one is not greater or lesser than the other, nor heavier, but simply different and what makes them alike is that they are conjunct in the group of planets of the Solar System, they are classified as planets of the solar system. For this, a common criterion has been determined for all of them, to revolve around the King Star. Likewise, for the attributes of things such as colors and hardness. If one thing is classified as red and another as red, then both are red. This seems a truism, but certainly both things may have slight differences in hue, both have the appearance of the color red. That is to say, we classify them as red, another thing is to see if one has more intensity of that color, but it would no longer be nominal. Here we are interested in getting to the classification

However, the interest of science at this level is not expressly to give them a name but to find a criterion that allows them to be grouped and classified. The classical logicians Cohen and Nagel (2000) maintain that

> In meaningful observation, we *interpret* what is immediately given to the senses. We *classify* the objects of perception (we call this a "tree", that a "star", etc.) by virtue of observed similarities between things, similarities that we consider meaningful because of the theories that we hold. (Authors' italics)

In other words, to classify means to sustain a hypothesis, we would not classify a whale as a fish because it has the characteristics of a mammal, therefore the senses are intelligent and come with a body of knowledge prior to observation.

Now, for quantification purposes, since what is observed and classified is the whole thing or a characteristic of it, without fractioning or weighing, we will have to assign it the number 1 and count the appearances of what is classified, which brings as a consequence that we will use that statistic that results only from the count of the appearances of what is observed. That is, the nominal is counted as a whole (as 1), e.g. a shirt, a house, a man, a dog, etc. or characteristics without stopping at the gradation of it; e.g. a red shirt, a wooden house, a mechanical man, a Coli breed dog, etc.

Let us now look at the scales that are constructed with the nominal or classificatory level.

Nominal Scales

By nominal scale we will understand the group of different names put together, grouped by means of a criterion. For example "planets that revolve around the Sun", another example will be "Capitals of Venezuela". Note that this scale needs a scale-forming, grouping criterion, which may be agreed upon in science or may be arbitrary by the researcher. We will see some nominal scales (some may be incomplete, in which case the reader will be able to complete them)

1. Household white goods

    1. Washing machine
    2. Dryer
    3. Refrigerator
    4. Blender
    5. Kitchen Assistant
    6. Toaster
    7. Microwave Oven

2. American car brands

    1. Ford
    2. Chrysler
    3. Dodge
    4. Chevrolet
    5. Jeep

3. Market Types

    1. Of money
    2. Tourism
    3. Capital

4. Total Market
5. Potential
6. Available
7. Target

4. Types of Flowers according to the ovary (taken from www.botanical-online.com

    1. Ovary Sulperus
    2. Semi-infertile Ovary
    3. Inferior

5. Dichotomous scales

    1. Yes/No
    2. True/False
    3. Presence/Absence
    4. Male/Female
    5. Male/Female
    6. Equal/Different
    7. Healthy/Sick

6. Colors

    1. White
    2. Black
    3. Blue
    4. Red
    5. etc.

7. Phobias

8. Syndromes

9. Organ diseases

10. Many other classifications. Innumerable classifications made by science

The purpose of the above list is not to exhaust, or even to show, the list of existing classifications, which is useless; I have only tried to exemplify the wide variety of classifications and to express that the important thing about this level of measurement is that the quantification of the elements of the families or groups is successful if the classification responds to a good hypothesis or classification criterion. For example, classifying animals as "blooded" or "non-blooded" may not be as useful if having or not having blood does not characterize animals better (having or not having blood is one more characteristic of being a vertebrate) than another more general criterion such as being a vertebrate

Ordinal Level. Comparative concepts

The size of the planets of the solar system (no longer the name) can be classified by the following set of expressions {very small, small, medium, large, very large}. These phrases are made up of adverbs (very) and adjectives (small, large, medium). Adverbs doing their job of modifying adjectives and giving meaning to quantity; "(Very) Small" is smaller in size than "Small" and this is smaller in size than "Medium".

Note that we use superlatives or diminutives as extremes of our individual perception, that is, the limit of such a comparison is given by superlatives and diminutives. Such superlatives are limits to express the perceived, so that if we wanted to express that something is more than "very much" we would have to resort to the psychological tone of the expression. The practical limits of this scale are the words with which we express the intensity of what we perceive

Now in this ordinal level we are referring to a characteristic of the planets (their size) and not to the planet as such. But we refer to that characteristic in degree (not as in the nominal level that we would say only if it has or not the characteristic), that is, which one has more of the characteristic than another.

The perception of the size of the planets and that grammatical structure form the instruments with which they are ordered. The ordering criterion can be simple agreement or (for other cases) standard estimates that demarcate the point of separation between what is considered "large" or "Very Large", "Low or "medium", and given that the observation has only perception and agreement as instruments, there cannot be external intervals or distances, mathematizable, fixable on paper or by other criteria such as "I believe that Jupiter is larger than Mars and Mars is larger than the Earth, do we agree? Now that these scales are treated as metric assuming a continuum of perception and standard units between one perceptions (small) and another (large) is another thing, many methods give good results without the isomorphism's being exact.

That is to say, this size is observed, catalogued or ordered according to the size of the planets, not according to indirect calculations of mass (even less direct). The ordinal is just that, an appreciation of the magnitude of something through some standard or convenient criterion (scientific or not). For example, a "small" person "A" is "small" for a bigger person "B", but this one can be "small" with respect to a bigger person "C". Then the order would be C > B > A. There is no specific measurement of weight (in kilograms and by means of a scale)

The set of symbols or numbers that we will use to denote such sorting will be any as long as it is satisfied that the order of assignment of the numbers corresponds to the real order of the objects under the sorting. E.g. If A> B it is because in reality what is denoted by A is greater than what is denoted by B. Using numbers it would be equal 2>1 if what is denoted by 2 is greater than what is denoted by 1. E.g. Fat = A and Thin = B, since as Fat> Thin, then A> B or if Fat = 2 then Thin = 1 since 2>1.

Finally, according to the given criterion, if 45 > 10, then Fat = 45 and Thin = 10. In other words, the numeration is still a code that only respects the order of magnitude, but there is no path from one rank to the other, there is no path between "Fat" and "Thin", there is no fixed interval, simply "fat is more than "Thin", let's say that one jumps from fat to thin and each jump is set by the observer.

Ordinal Scales

We will call ordinal scale to the set of nouns, polar or not (generally adjectives), ordered by means of gradations of the same, for example High, Medium and Low or, Always, Almost Always, Never.

Let's see some ordinal scales written in two alternatives but extensible to even or odd scales of more levels, bidirectional or unidirectional. For example, the attitude scale expressed by means of the adjectives Bueno and Malo can maximize its intensity by placing the superlatives Buenísimo (or bonísimo) and Malísimo at the extremes and follow the degradation, thus, Buenísimo, Muy Bueno, Bueno, Bueno, Ni Bueno Ni Malo (midpoint of the scale), Malo, Muy Malo and Malísimo. We will not dwell on scale construction strategies, but we wish to show some of them so that the reader becomes familiar with them. The dichotomous (two-level) scales can be extended as indicated above by using adverbs of quantity and the augmentatives or diminutives of the expressions

1. Distance
    1.1. Nearby
    1.2. Far
2. Aesthetics
    2.1. Beautiful
    2.2. Ugly
3. Attitude
    3.1. Good
    3.2. Regular
    3.3. Bad
4. Temperature
    4.1. Cold
    4.2. Hot
5. Contingency
    5.1. Probable
    5.2. Unlikely
6. Contingency

6.1. Insurance
6.2. Uninsured
7. Frecuency
    7.1. Always
    7.2. Never
8. Acceptance
    8.1. Very much
    8.2. Not at all
9. Height
    9.1. Tall
    9.2. Low
10. Expectation
    10.1. Favorable
    10.2. Unfavorable
11. Sesitivity
    11.1. Sensitive
    11.2. Insensitive

12. Agreed
    12.1. Agree
    12.2. Disagree
13. Fairness
    13.1. Fair
    13.2. Unfair
14. Quality
    14.1. Excellent
    14.2. Bad
15. Temporal Effectiveness
    15.1. Early
    15.2. Delayed
16. Speed
    16.1. Fast
    16.2. Slow

Metric Level. Metric concepts

Planets have mass and therefore weight. But no longer perceptually but by indirect measurements (the weight of Mars could not be measured directly), the weight can be determined by mathematical formulations that approximate it. In another case, the weight of people can be determined directly by means of a scale. The metric, in general, is determined by means of instruments external to the human being that respond to scientific criteria, reference systems that do not depend on the researcher but on international conventions. These conventions are criteria on laws of nature that standardize or regulate the behavior of variables beyond the will and conscience of the researcher. Measurements such as the meter, the kilo or the temperature respond to conventions or laws of behavior that are concretized in physical instruments. Therefore, everything that can be compared against external and universal reference systems such as those created for the International System of Weights and Measures (SI) has a metric. For example, the BIT, the kilo, the meter, degrees Celsius, etc. are metric units that measure extensive or intensive characteristics of matter.

In the social, psychological or medical sciences, many characteristics are measured by means of such instruments, but in fields where will or intentionality are involved, metrics have been of little help, although nominal and ordinal variables have been.

Chapter 2. Quantifying Events and Phenomena

Regularities and Frequencies

Regularities

Statistics begins by counting frequencies and relies on them to develop its techniques; therefore, we need to know them well, their types and ways of calculating. Frequency, in the perspective of everyday life, means a contact with events of reality whose repetition helps us to know them and make decisions. If something is repeated several times, it gives us greater certainty in the belief that the next time it will happen again if the circumstances are similar; if something were to stop happening it would influence a negative belief about the next occurrence. This is known as the problem of induction and it has always been discussed whether such a form of contact with reality can give rise to scientific knowledge. However, although epistemologists have discarded it because the existence of an *Inductive Principle* that supports such a form of knowledge or inference cannot be proved, we see that it continues to be used in science as a secondary inference, as an inductive support for the decisions that are taken and in everyday life as a way of generalizing beliefs that we have; this process of generalization in everyday life can be detrimental, as there is nothing to indicate that such behaviors cannot change, yet we live daily subject to this risk by acting on the basis of close or apprehended experience.

Science uses the observation of cases not only inductively but also accompanied by theories that support the scientists' judgments. Eloquent are the cases of the reappearance of the symptoms of a disease in a medical post-treatment; the psychological knowledge of people based on the recurrence of behaviors; the prediction of a winner in an electoral race based on the greater recurrence of favorable mentions; the increasingly recurrent appearance of certain birds as an indicator of climate change; etc. These examples and many others show the uses we make of recurrent events to decide on a course of action. However, the mere fact of observing the recurrence is not a determining factor in forecasting, since we may wait and nothing happens. But there is no doubt that the observation of reality, accompanied by hypotheses or well-founded theories, is a source of information that we use every day, and science uses it in a prudent manner.

Nowadays, *epistemology* has classified these regularities by giving them a better profile that allows us to know their origin. Four types of *regularities* are known, which according to Díez and Moulines (1999, p. 129) are synthetic of all types and which include our statistical frequencies. 129) are synthetic of all types and which include our statistical frequencies, we will describe them quickly without further discussion; we have *the analytical or conceptual regularities*, which refer to the way we know things once the concepts have been fixed; for example, when we say "rational animals are animals" or "bachelors are not married" we are applying two concepts on events of reality, namely, that of "rational animal" and that of "bachelor", and we infer, deductively, that a rational animal is an animal and that a single person is not married, according to our concepts and they are regularities by the repetition we make of the judgment or concept. Then we have *nomic regularities*, which refer to natural laws, such as the law of gravity and in general, all the laws of the sciences of nature of which repetitions are expected with rare exceptions. Thirdly, we have *the factual or accidental regularities*, which refer to merely occurring events, that is, events that simply occur but might not have occurred; they occur in fact, but not in law; for example, certain traffic accidents could have been avoided if other things had happened (such as having foreseen the danger, etc.); they are events for which it is very difficult to give certain reasons for their occurrence. Finally, we have *epistemic regularities* that are anthropomorphic regularities; that is, they do not exist in fact in nature, but are constructed by human beings; for example, we know crows as "black" although there is nothing in nature that prevents them from being white; likewise with zebras, we know them with black (or white?) stripes but they could be without stripes. This type of regularity is anthropomorphic in the sense that being black or having stripes is something set by the human; that is, even if there could be zebras without stripes and white crows we know them respectively, with stripes and black. In the same way we have the deontological rules that are rules that are not in society, but are imposed by the guilds; thus, the legal laws are social constructions that when obeyed cause regularities or behaviors that conform to the law or codes of ethics.

Statistics does not discriminate about these types of regularities; they are present in the studies, going unnoticed or, simply, being taken as evident depending on the discipline that uses them. For example, we construct nominal scales of marital status, presupposing the conceptual or epistemic regularity of single, married, etc., although the researcher fixes his concepts when he is going to carry out the study; for example, a psychologist can determine different types of age range for adolescents, according to the theories involved in his study or the country where it is carried out; today items on sexuality are accepted that were not admitted before. Regarding nomic regularities, statistics evaluates the way in which events occur in time and space; for example, atmospheric pressure is a nomic regularity that does not occur in the same way according to certain initial conditions; it is not the same atmospheric pressure on the beach as on a very high mountain; likewise, the prevalence of certain symptoms of a certain disease may be determined by skin type or gender. Regarding factual or accidental regularities, science resorts to probabilities, since there are events that simply occur and although they may be supported by some theory, their recurrence is very uncertain, such as earth tremors or earthquakes. Finally, epistemic regularities are studied according to the degree of compliance with rules; for example, people who obey traffic rules, or those who arrive at work on time. As we can see, we have used all these types of regularities in statistics without being aware of the classification used in Epistemology. In general, in order not to cause cognitive dissonance, we will leave the term "regularity" for epistemology and use our common term "frequency" to refer to the act of counting or inferring a certain quantity, but we should not lose sight of that typology of regularity because it could be very useful when we need to draw conclusions. Therefore, we will go on to define *frequency* and make a discussion about them.

Frequency and its distribution

The work in statistics begins with the counting of cases, counts or frequencies. We will say that **a frequency is the appearance of a value in the sample**, for example, let us suppose that the values of the ages that appear in a sample are {21, 35, 18, 35, 35, 48, 11, 39, 40, 40, 35}, each one of the ages that appear in the sample is a frequency; that is, age 21 has a frequency, age 48 has a frequency; age 40 has two frequencies, age 35 has three frequencies, etc. If it appears only once, it will be one frequency, if the same value appears twice, it will be two frequencies, and so on.

By frequency distribution we mean the number of frequencies that appear of each value and can be written in several ways, in tables or graphically. For the previous case we will have the following frequencies (without ordering the values) (see table 1)

Tabla 1 Ejemplo de formación de frecuencias

Age value	21	35	18	48	11	39	40
Frequency or No. of times repeated	1	3	1	1	1	1	2

The first row is the values being studied, i.e. the ages, and the second row is the frequencies or **frequency distribution**. Age 21 appears once, age 35 three times, etc. Statistics is based on the number of times that one or more values are repeated.

Tabla 2 Example of a nominal frequency distribution

		N	%
Municipality of origin	1. Liberator	58	56,9
	2. Chacao	2	2
	3. Baruta	15	14,7
	4. Other place	27	26,5
Group total		102	100

Let's look at an example, table 2 shows the municipality of origin of 102 people attending a shopping mall. The category title is "Municipality of Origin", the category labels are the names of the municipalities (note that they are coded), the scale is nominal since these are names of municipalities, the labels of the statistics are N and %.

The distribution of frequencies are the two columns (shaded) that comprise the body of the table (N and %), the distribution of the count (N) we call absolute frequencies and the distribution of the percentage (%), relative frequencies. They are absolute frequencies because the count of the frequencies was direct and the value is not being compared with any other, i.e. the 58 values of the municipality of liberador are raw counts. On the other hand, the relative frequencies are so called because the value of N has been relativized, compared against the total. In other words, 56.9% is the result of dividing (comparing) 58/102 and multiplying it by 100. It reads "56.9% of the total (relative) number of people attending the mall come from the municipality of Libertador". The same would be true for an ordinal variable

Tabla 3 Example of ordinal frequency distribution

	Enrollment of the first four years of secondary education				
	First	Second	Third	Fourth	Total
N	78	63	55	21	217
%	35,94	29,03	25,35	9,68	100

In the example in Table 3 we see a distribution of ordinal frequencies. In this case the distribution is organized horizontally (the previous one was vertical). The shading in the table shows the distribution.

Note that for both cases the frequency is constructed by counting the units.

The Unist

An important aspect of frequencies is the units in which they measure objects.

If I write the sign "5", the pure digraph or sign means nothing. I can *intrterpret* that it is the "Arabic" five or number 5; but such a sign could be written as "V", that is to say a high V, which would be the "Roman" five. That is, the sign does not tell me anything, it is only nominal and acquires *mathematical significance* within a scheme of logical rules (Arabic or Roman numerals). But, even so, that "5" if it is not connected with reality in the process of measurement does not acquire real, empirical significance. For example, that 5 may be the result of

1. Count five objects. 5 shirts
2. Measure five liters of a liquid. 5l.
3. Weighing five kilograms of a mass. 5k.
4. To measure temperature in degrees Celsius. 5°C
5. Measure the capacity of a computer memory in gigabytes. 5 GB.

As can be seen, the value accompanied by the unit of measurement acquires empirical meaning (it is no longer "any five" but five units of something) and now it can be *compared* with another beyond a mathematical comparison, an empirical comparison. Pe. 10 shirts minus 5 shirts leaves another 5 shirts, etc. But 5 shirts minus 3 oranges would be empirically meaningless. It would be an empty concept (set)

Ratios and Rates as Frequencies

Ratios as Rates

We saw that absolute values as quantities representing intentional counts of objects or events are meaningless if they are not compared against other quantities, the reason is simple dual linguistic expressions such as hard, soft, ugly or beautiful are difficult to compare because we always need to know how to compare and order things, for example we would not know how to say which of two "tall" people is taller unless we measure them; but there are more complex concepts such as diseases whose absolute values are useless if they are not compared, for example, against the number of inhabitants. On the other hand, it is not always possible to make direct counts but through partial operations and then total them, so that frequencies can be known in different ways, directly or in a derived way by relativizing them using quotients, sums or differences. Let us briefly look at these aspects

Ratios compare quantities having the same or different units

Example 1 If there are 25 men and 34 women, what is the ratio between them? Here we have different units, we can compare men with respect to women, so (25 men)/(34 women)=0.735 men/woman and read 25 men for 34 women is equivalent to 0.735 men for one woman.

Example 2 How much do the 23 workers in the office represent with respect to the 130 workers in the company? Note that we are comparing quantities that have the same units, no. of workers. In the operation, the units are cancelled in this way (23 workers)/(130 workers)=0.1769. We read, with respect to one worker in the company, the one in the office represents 0.1769 or 0.1769/1; 0.1769 out of 1

It should be noted that when the ratio compares quantities of different units it is called a **Rate**. The first example is a rate. It reads the ratio of males to Females is 0.735.

To recapitulate, ratios compare two quantities of similar or different nature, when they are of different nature it is called rate

Proportions

Ratios are quotients where the numerator is contained in the denominator, for example, if there are 25,000 inhabitants in a population and 1,500 are diagnosed with diabetes, the proportion of inhabitants with diabetes will be

$$\frac{1.500 diabetics}{25.000 population} = 0,06 diabétics/population$$

Reads 0.06 diabetics per capita.

Note that the values being compared have different units, which makes the proportion a ratio, but with the particularity that the denominator contains the numerator, i.e. we are talking about similar things, of the same nature.

A characteristic of proportions is that since the numerator is contained in the denominator, the numerator will always be smaller or equal to the denominator, therefore, the resulting fraction or quotient will always be less than or equal to the unit. Let us look at a general example. If a+b = c, the proportion of a and b over c are

$$\frac{a}{c} = \frac{a}{a+b} \quad y \quad \frac{b}{c} = \frac{b}{a+b}$$

The sum of both ratios is equal to one. That is to say

$$\frac{a}{a+b} + \frac{b}{a+b} = 1$$

In the example of diabetics, if there are 1,500 diabetics out of 25,000 inhabitants, it indicates that the remaining 23,500 (25,000 - 1,500) are not diabetic, therefore

$$\frac{1.500 diab}{25.000 habit} + \frac{23.500 no.diab}{25.000 habi}$$

$$= \frac{0,06 diab}{habit} + \frac{0,94 no.diab}{habit}$$

$$\frac{(0,06 diab + 0,94 no.diab)}{habit} = 1$$

To recapitulate, ratios compare two quantities of similar or different nature, when they are of different nature it is called rate, when the denominator of the ratio contains the numerator, it is called proportion. Percentages are rescaled ratios with base 100 although the base of comparison can be greater in multipliers of 10.

Tipos de Frecuencias

Based on the above, we can classify the frequencies as shown in diagram 1.

Diagrama 1 Classification of frequencies

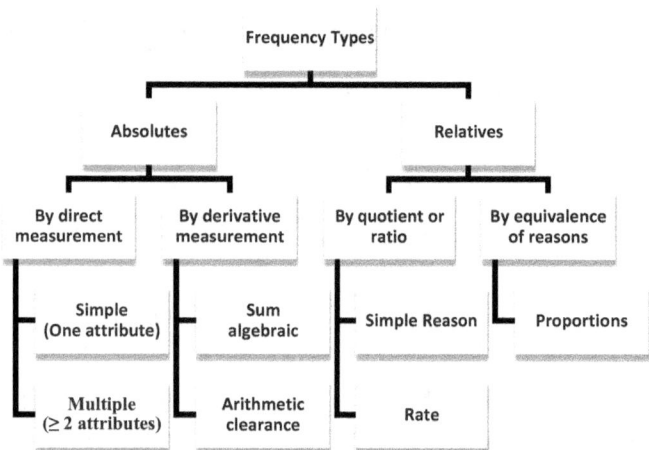

Absolute frequencies are divided according to whether the measurement or counting of frequencies is *direct* or *derived*. When the count is direct, frequencies can be single or multiple; *single* when the values of one category or attribute are counted (e.g. male or Female) and multiple when the combined values of two or more categories or attributes are counted (e.g. male and healthy; male, healthy and over 18 years old, etc.) [4].

Frequency is derived when it is inferred by means of an *algebraic sum* (includes equal or opposite signs) or by quotient; for example, it is inferred that the total hours of class (*THC*) is equal to the partial totals of the hours taught for subjects a, b and c (*THCa,THCb,THCc*). So (*THC=THCa+THCb+THCc*). Now, we have frequencies by *arithmetic clearance* when we have equations formed by sums and we clear some variable knowing the others. For example, the final stock of a product (*Ef*) in a warehouse plus Sales (*V*) is equal to the initial stock (*Ei*) of the product plus Purchases (*C*). That is (*Ef+V=Ei+C*). If we want to know the *Ef* we measure the other three variables and clear *Ef*. In all these cases we obtain absolute frequencies by absolute addition or subtraction.

However, if we establish quotients between values, the result will no longer be absolute due to the effect of the division, it will be a relativized value and we will have relative frequencies. Thus, we have frequencies by quotient (or ratio), as for example when dividing the number of men (M) with respect to the number of women (W) we obtain the number of men for each woman. E.g., if we have 15 men and 5 women, then we will have, $\left(\dfrac{15\,M}{5\,W} = 3\dfrac{M}{W}\right)$ 3 men for each woman. It is a frequency constructed "with respect" to another figure. Another type of relative frequency is that which is obtained from an equality of quotients or ratios to establish the proportions

This whole subject gives rise to the Indicators that are expressed by means of ratios, proportions and rates; we are particularly interested in this work in the frequencies, both absolute and relative, which are obtained directly to construct statistical tables, although the rest of the indicators can be expressed in tables. That is to say, we will see the case of the construction of tables from counts on lists or databases and a particular case of the types of responses in questionnaires, single and multiple responses.

---

[4] In Logic it is said that intension or connotation is added (different from intentionality).

Particular cases of frequencies

Single and multiple absolute frequencies

Thus, there are single-answer and multiple-answer questions. The processing of the simple answer is the best known, it is the usual relative frequencies of the tables (division of the absolute frequency by the total). For example, those dichotomous options (e.g. Yes/No) or polytomous options (Yes/ No/ I don't know; White/Black/Blue, etc.) in which the answer to one option logically excludes the others.

But we have multiple responses, for example, in the question of possession of white and brown goods, we ask if the respondent has a refrigerator, stove and television; the respondent can say that he/she has one of the items, two or all three. This generates two bases of comparison, the first comparing responses over total cases and the second over total responses. The first case is a ratio, the second a rate. Let's look at this in more detail, starting with single frequencies and then with multiple and finally relative frequencies. The following list shows frequencies of gender, age and height for 30 people.

Tabla 4 Some values for gender, age and height

Gender	Age	Height
1	18	1
2	36	2
1	30	3
2	35	2
2	30	3
2	35	2
1	21	1
2	22	1
2	27	2
2	22	3
1	20	2
2	37	3
2	30	3
2	20	3
1	22	2
1	20	1
1	32	2
1	19	3
1	23	2
1	36	3
2	31	2
2	27	3
1	36	2
1	20	3
1	28	3
1	35	3
2	38	2
2	21	1
2	27	1

2	27	1

Given these variables we will say that **simple absolute frequencies are those that are obtained from each variable without combining with others** and we will say that **double absolute frequencies are those that are obtained from the combination of the frequencies of two variables**. It is possible to speak of triple, quadruple, etc. frequencies. A particular case of triple frequencies are the pivot tables that we will see below.

The table above contains values for age in years, gender coded as {1= Male; 2= Female} and height code measured ordinally with codes {1= Short, 2= Medium, 3= Tall} in 30 persons.

Calculation of simple absolute frequencies

We ask ourselves how many males and Females are in the sample? Or what is the gender distribution in the sample, since the male gender code is 1 and the Female gender code is 2, we count the 1's and 2's resulting in 14 males and 16 Females. We will say that **the gender distribution in the sample is 14 males and 16 Females**. These counted frequencies are simple absolute or simple frequencies because they were taken from the gender variable without the other variables age and height intervening. It is also said that gender is taken independently of age and height.

Similarly, we can take the simple frequencies of all ages or of one in particular. Let's calculate the simple frequency of 20 years. To do this we count the ages 20 resulting in 4 frequencies. We say that there are 4 frequencies of 20 years. Again these frequencies are simple because the other two variables did not intervene.

Calculation of double absolute frequencies

Double frequencies consist of counting the values of two variables, for example, suppose we want to process the variables, gender and housing tenure. Suppose further that the labels and codes for gender are {1= male, 2= Female} and that those for housing tenure are {1= has, 2= does not have}, if we wanted simple gender frequencies we would count the 1 for male or the 2 for Female, but suppose we wanted to know how many people are male and have housing, obviously we would have to count the codes for male gender and has housing, i.e. the pairs 1 and 1. If we wanted to know the number of people of the male gender who do not have housing, we would count codes 1 and 2, and so on.

Example 1 For the example in the table above, let us consider

1) Men of short stature,
2) Women of short stature
3) Men of 18 years of age
4) Men of medium height who are 30 years old.
5) Women between 18 and 25 years of age.

Solution.

1) Given that the male gender code is 1 and the short stature code is 1, we must look for the combination {1,1} in the first and third columns, resulting in 3 short men.

2) Given that the Female gender code is 2 and the short stature code is 1, we look for the combination {1,2} resulting in 4 short women.

3) Males of 18 years old are 1

4) Medium height men who are 30 years old are 0

5) Women between 18 and 25 years old are 4.

Relative frequencies
Simple Relatives

The simple relative frequencies are obtained from the simple absolute frequencies. In the above case of gender distribution in 14 males and 16 Females out of a total of 30 persons the simple frequency for males is (14/30)*100 = 46.67% and for Females (16/30)*100 = 53.3%. Note that the sum of the percentages gives 100, which is obvious since the alternatives are logically exclusive. These percentages are the most common, but those resulting from multiple responses are poorly understood.

Multiple relative

The case of multiple relative frequencies is little treated, although it is more useful, the processing of data from the so-called multiple response is little known. Many studies do not apply them because of ignorance of the procedure or because of the difficulty of interpreting percentages that add up to more than 100. The widespread belief that percentages should give only 100 is very strong, but is unfounded, since this depends, again, on the basis for comparison.

This is the case of those nominal scales where the respondent can mark more than one (or all) answers. For example, in a hypothetical interview, 22 people are asked if they have any, all or none of these items: 1= refrigerator, 2= stove and 3= television. Table 5 shows the 23 hypothetical responses.

Note that in case 1 the person has all three items (1= refrigerator, 2= stove and 3= television) and that person 5 does not have a refrigerator. This type of multiple response generates two bases of comparison, one on the total number of cases (of respondents) and two on the total number of responses (items reported), thus, two types of percentages appear, namely, percentage of cases and percentage of responses.

Table 5. **22 interviews on refrigerator, stove or television ownership**

Case	Refrigerator	Kitchen	Television	Case	Refrigerator	Kitchen	Television
1	1	2	3	12	0	2	3
2	1	2	0	13	0	2	3
3	1	2	3	14	1	2	3
4	1	2	0	15	1	2	3
5	0	2	3	16	1	0	3

6	0	2	3	17	0	2	3
7	0	2	3	18	1	2	0
8	1	2	3	19	1	2	3
9	1	2	0	20	1	2	3
10	1	0	3	21	1	2	3
11	1	0	3	22	1	0	3

How is this information processed, and how is it interpreted? Table 6 shows this simple procedure. Of the 22 people interviewed, 16 have a refrigerator, 18 have a stove, and the same number have a television, giving a total of 52 responses. That is, 22 people (cases) gave 52 answers, these are the two bases of comparison. Now, if we take the 16 refrigerator frequencies against the 52 responses (base) (sum of the refrigerator, kitchen and television responses) times one hundred, we get 30.8%, indicating that of the total number of items that the 22 people have, 30.8% of the items are refrigerators. On the other hand, if we take the 16 refrigerator frequencies against the 22 cases (interviewed) per hundred, we get 72.7%, indicating that 72.7% of the 22 people interviewed have a refrigerator. In other words, of the 52 items owned by the 22 people, 30.8% are refrigerators and of the 22 people, 72.7% have a refrigerator (see Table 6).

Table 6 Percentage of responses and cases of ownership of refrigerator, stove, or television set or television

Item	Frequency	% Base 52 responses	% Base 22 cases
Refrigerator	16	30,8	72,7
Kitchen	18	34,6	81,8
Television	18	34,6	81,8
Total	52	100	236,4

Now, what is confusing is that 72.7% of people have a refrigerator, 81.8% have a stove and another 81.8% have a television, and that this adds up to 236.4%. Let us point out some aspects

1. The fact that the percentage of cases is higher than the percentage of responses only indicates that some people have more than one item, since if the responses coincided with the cases, then they would have only one.

2. Keep in mind that the frequencies counted are not the people but their responses. When the question is a single-choice question, the number of responses coincides with the number of people, but when the response is multiple, this is not necessarily the case; generally the number of responses is greater.

3. The responses of the persons are analyzed.

4. Another characteristic is that, in the simple response, given that the options are exclusive, the percentage of favorable responses to one option determines the others (since they are its complement). For example, if 16 of the 52 items held by the 22 people are a refrigerator and 18 are a kitchen, then logically the remaining 18 are a television. But in the multiple response, the processing of the options is not

independent, i.e., 16 have a refrigerator, but it does not exclude that they have a kitchen or a television, so that the intersections (double and triple frequencies) are then taken into account.

5. That is, of the 16 people who have a refrigerator, 4 have a kitchen, 4 have a television and 8 have all three. This breaks down the percentage of those who have a refrigerator (72.73%) into 18.18% (refrigerator and kitchen); another 18.18% (refrigerator and television) and 36.36% all (Table 7 shows this fact). The same is true for those who answered kitchen and television (check the other percentages).

Table 7. Holdings of broken down items

	Refrigerator	Kitchen	Tele	Tres	Total
Refrigerator		4	4	8	16
% fridge		18,18	18,18	36,36	72,73
Kitchen	4		6	8	18
%cook	18,18		27,27	36,36	81,82
TV	4	6		8	18
% tv	18,18	27,27		36,36	81,82
				Total responses	52
				Total Percentaje	236,36

Conversion of quantitative scales to ordinal scales using the clustering technique

Data clustering is a classic technique in statistics. Traditional statistical texts began the description of methods with the process of data clustering. This technique consisted of reducing the volume of information to ranges or intervals called **class intervals**. The method was necessary due to the lack of automated processing that made it impossible to process large volumes of data manually. Performing a statistical operation meant counting and recounting very large columns of data, so reducing the mass of data to a few intervals reduced processing considerably. Nowadays computers have great processing capacity and speed, which makes it useless for statistical analysis to start with the grouping of data.

However, the technique of grouping data remains valid for the case of **reducing quantitative variables to qualitative** variables for better understanding. For example, data on arterial hypertension are more comprehensible when reduced to an ordinal scale of Low, Medium and High; grades to a scale of Good, Fair and Poor or another broader scale.

Data clustering method

An example will suffice to know the method. Suppose we have the following values {10, 15, 11, 14, 14, 14, 16, 12, 11, 11, 14, 11, 11, 11, 11, 8, 18, 11, 14} and we need to group them five by five. For that we have to create a class or interval with values ranging from 0 to 5; another with values from 6 to 10; another from 11 to 15 and another with values from 16 to 20. Then we count how many values fall into the chosen ranges and the results are shown in table 8 below

Table 8 Grouping of data into 4 classes and their frequencies

Ordinal Distribution of Frequencies	
Class or Range	Frequency
De 0 a 5	0
De 6 a 10	1
De 11 a 15	12
De 16 a 20	2
Total	15

Let's look at some characteristics of this data grouping process.

1. When grouping the data, the individuality of the data is lost, that is, in the interval from 6 to 10 there is a frequency as can be seen, but we do not know which value it corresponds to, we do not know if it is a 6 or an 8, etc.; similarly, the interval from 16 to 20 has 2 frequencies, but we do not know which values it corresponds to. For this reason the scale created (the classes) is ordinal. We have transformed a quantitative scale into a qualitative one.

2. The sum of the frequencies corresponds to the original total of values.

3. The size of the classes can be arbitrary or determined by limits established by international conventions. For example, one researcher may consider childhood to be a period between 1 and 12 years, but another may consider it to be between 1 and 10 years. Another example is the classification of the amount of Triglycerides that according to the American Herat Association would be in the following order

(In mg/dL)

Normal range, low risk <160
Near high range 150-199
High: at risk 200-499
Very high: High risk >500
   (En mg/dL)

Ejercices

1.   Table 9 below shows 30 values for age, sex, weight, height, cholesterol and triglycerides from a health survey. With this information find

   1.1.   To code the variable Sex

   1.2.   To find the simple frequencies of sex.

   1.3.   To find the simple frequencies of age

   1.4.   To find the simple frequencies of triglycerides (Trig)

   1.5.   To find the simple frequencies of cholesterol (Chol)

   1.6.   To find the double frequencies of male and even ages

1.7.	Find the double frequencies for Female and odd ages.

1.8.	Find the even-aged and odd-aged double frequencies.

1.9.	Find the even-age and even-triglyceride double frequencies.

1.10.	Group the ages in the following groups [under 40; between 41 and 50; over 51] 1.11. Group the sizes in the following groups [under 40; between 41 and 50; over 51].

1.11.	Group the sizes into low (from the smallest to 1.5), medium (from 1.51 to 1.65) and high (from 1.66 to the largest).

1.12.	Group the weight in the following groups [from the smallest to 50.0; from 50.1 to 70.0 and from 70.1 to the largest].

1.13.	Group triglycerides using the following scale: Normal: less than 150 ; g/dL; Borderline high: 150 to 199 mg/dL; High: 200 to 499 mg/dL; Very high: 500 mg/dL or higher.

1.14.	Group cholesterol using the following scale: Desirable level < 200; Borderline high 200 to 239; High ≥ 240.

1.15.	Find the double frequencies of normal triglyceride and ages less than 40.

1.16.	Find the dual frequencies of male and cholesterol desirable level.

1.17.	Find the double frequencies of Female and high cholesterol

1.18.	Find the double frequencies of ages between 41 and 50 and weights less than 50.

1.19.	Find the double frequencies of age between 41 and 50 and borderline high triglycerides.

1.20.	Find the double frequencies of male and borderline high cholesterol.

1.21.	Find the double frequencies of median size and weight between 50.1 and 70.

Table 9 Values for age, sex, weight, height, cholesterol and triglycerides for 30 individuals

N°	Age	Gender	Weight	Height	Col	Trig	N°	Age	Gender	Weight	Height	Col	Trig
1	36	1	73	1,51	254	300	16	48	1	79,4	1,58	237	151
2	63	1	59,2	1,49	208	410	17	67	2	65,4	1,62	175	94
3	42	1	73,6	1,47	232	201	18	43	1	59,4	1,53	212	88
4	69	1	53,6	1,38	166	257	19	45	1	57,3	1,56	280	205
5	43	1	68,2	1,43	232	200	20	53	1	66,3	1,48	233	160
6	55	1	62,3	1,55	200	180	21	59	1	95,7	1,53	273	150
7	59	2	55,6	1,7	220	130	22	55	2	71,5	1,63	315	210
8	49	2	80,4	1,66	276	253	23	39	1	65,3	1,55	263	117
9	60	1	74	1,46	280	186	24	68	1	61	1,48	206	109
10	43	1	68	1,5	267	423	25	48	1	64,5	1,55	288	210
11	50	1	66,5	1,47	180	160	26	58	1	60,2	1,52	309	147
12	65	2	87,8	1,72	161	200	27	37	1	102,5	1,6	229	270
13	54	2	89,6	1,63	230	300	28	50	1	82,5	1,5	214	122
14	48	2	75,9	1,69	250	180	29	49	1	70,9	1,6	212	167
15	38	1	60,3	1,51	180	150	30	47	2	80,2	1,68	281	317

2. In a Venezuelan city, a manufacturer of pasteurized products conducted a survey of 97 people to test the most important brands of liquid yogurt (Nutrigurt, Frutigurt, Alpina and Yoplait) that they consume. Respondents could check one or all of the test brands and these were the answers; Nutrigurt =91, Frutigurt =20, Alpina =25 and Yoplait =20. Get

1. Percentages on cases

2. Percentages of responses

Note: Refer to Table 6

3. The following database (Table 10) shows 40 cases of a survey of teachers at a regional university in Venezuela (UNET) on the most important characteristics that the Rector of the University should have. The 40 teachers were asked to rank 4 options (1= Firm, 2= Efficient, 3= Planner, 4= Practical) according to their importance. For example, teacher 1 selected for the first option to be a planner, then efficient, then practical and finally firm. With this information

   1. Obtain for the first preference the simple frequencies of the four options.

   2. Obtain for the second preference the simple frequencies of the four options.

   3. Compare the simple relative frequencies of the previous cases and say which characteristic stands out in both?

   4. Repeat 1, 2 and 3 for the third and fourth columns.

   5. Compare the simple relative frequencies obtained and order them, what is the order of the characteristics that the Rector should have?

   6. How many cases are there in the database?

   7. How many responses are there in the database?

   8. Now analyze the database as a multiple response and obtain the percentage of cases and responses.

   9. Compare the results obtained in item 8 with item 3, what can you say?

Table 10 Responses from 40 teachers on the desired attributes of the Rector of UNET

	First	Second	Third	Fourth
1	3 planner	2 efficient	4 practical	1 firm
2	1 firm	3 planner	4 practical	2 efficient
3	3 planner	4 practical	2 efficient	1 firm
4	4 practical	2 efficient	3 planner	1 firm
5	3 planner	2 efficient	4 practical	1 firm
6	2 efficient	1 firm	4 practical	3 planner
7	3 planner	4 practical	2 efficient	1 firm
8	2 efficient	3 planner	4 practical	1 firm
9	3 planner	1 firm	4 practical	2 efficient
10	4 practical	2 efficient	1 firm	3 planner
11	4 practical	1 firm	3 planner	2 efficient
12	2 efficient	3 planner	4 practical	1 firm
13	3 planner	4 practical	2 efficient	1 firm
14	3 planner	2 efficient	4 practical	1 firm
15	4 practical	1 firm	3 planner	2 efficient
16	1 firm	4 practical	2 efficient	3 planner
17	0	1 firm	0	2 efficient
18	4 practical	3 planner	1 firm	2 efficient
19	2 efficient	1 firm	3 planner	4 practical

	First	Second	Third	Fourth
20	3 planner	2 efficient	4 practical	1 firm
21	2 efficient	1 firm	3 planner	4 practical
22	4 practical	1 firm	3 planner	2 efficient
23	3 planner	1 firm	4 practical	2 efficient
24	2 efficient	3 planner	1 firm	4 practical
25	1 firm	3 planner	2 efficient	4 practical
26	3 planner	4 practical	1 firm	2 efficient
27	4 practical	3 planner	1 firm	2 efficient
28	1 firm	3 planner	2 efficient	4 practical
29	2 efficient	3 planner	1 firm	4 practical
30	3 planner	2 efficient	1 firm	4 practical
31	3 planner	2 efficient	1 firm	4 practical
32	2 efficient	4 practical	1 firm	3 planner
33	2 efficient	1 firm	4 practical	3 planner
34	1 firm	4 practical	2 efficient	3 planner
35	2 efficient	3 planner	4 practical	1 firm
36	1 firm	4 practical	3 planner	2 efficient
37	3 planner	2 efficient	4 practical	1 firm
38	4 practical	3 planner	2 efficient	1 firm
39	2 efficient	3 planner	4 practical	1 firm
40	3 planner	2 efficient	4 practical	1 firm

PART TWO

Tabular Structure

Statistical table and usability

Statistics and Data Processing go hand in hand, the data of a survey, census or administrative record come in the form of questions or items that are answered or filled out by people, these elements for processing change their form from a spreadsheet or questionnaire to a tabular form of lists or databases where individuals or cases are placed in the rows and in the columns the variables or items and from there comes the statistical process of applying formulas of means, correlations, variances, sums, divisions, among many other processing results that we call summary statistics and that are conveniently placed in a statistical table or processing summary table, correlations, variances, sums, divisions, among many other processing results that we call summary statistics and that are conveniently placed in the statistical table or processing summary table. However, this table is not a simple sanctuary where data are displayed because they must be read or interpreted according to a research plan and generally these two processes are divorced, data processing offers a summary table and the researcher or teacher often does not know how to read the data because the table has been presented as a data display starting the usability problems of statistical tables. The simple ones of one variable are easy to read, but the larger and more complex ones cause visual chaos because the first thing you see are a lot of incomprehensible values and percentages, on the other hand the manuals of the statistical programs do not offer definitions or methods that solve the problems of usability of the statistical tables although they teach how to elaborate them from their windows, these manuals tell the user how to place the variables in some position and even how to combine it with another and which statistics can be used and in which position it should go according to the user's criteria. This is enough because it teaches how to use the tables module, but the success is relatively little because there is no methodological interface that allows the full understanding of the use of tables, one thing is that a user learns to build tables and another that he can transmit this knowledge outside the respective software.

This second part shows the structural elements of a statistical table, in the third part we will see its logic and how the table is dynamized. The usability of the statistical table will depend on the verification of all the constituent elements of the structure, i.e. a statistical table must be auditable. We will define two large blocks of verifiable elements.

Usability

Table structure
The table must contain all the internal and external elements properly defined, i.e. the title must correspond to the content, the variables must contain the items properly placed, the data source must be verifiable and fully described, i.e. there must be no doubts about its internal and external elements.

Reading order of the values in the table

Contrary to other cultures, in our culture the reading is made from left to right and from top to bottom, therefore the labels that guide the reading must be placed on the left to be read together with the statistics (we will see this later), likewise the values must have the references of their calculations, that is, the percentages must have their bases of comparison, the summed values must have their totals, since these indicate the direction of the reading. Often the formulas that allowed the calculations are placed in the table footnotes, especially when there are estimates of figures that do not come from the databases. In the third part we will see that from the analysis of the semantics a table can be planned.

Chapter 3. Essential Elements of a Board

First we will define the elements necessary to form a table, i.e. a grid (the basis of the table), the categories to be shown (properties or variables) and the measurements or numbers that differentiate the categories. Then we will see the components of a table, its types and the techniques to build tables.

Matrix definition of table and its description

Description of a Grid

Ilustration 1 rid of 8 rows, 5 columns and 40 cells.

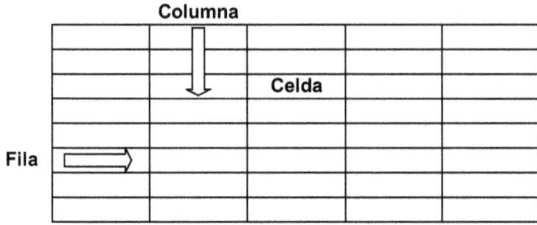

A grid is a grid formed by the intersection of rows and columns. The rows are the horizontal entries and the columns are the vertical entries of the grid. Figure 1 shows a grid consisting of 8 rows and 5 columns. The grid is the basis from which the table is made.

The intersection of the rows and columns of the table produces a space called a *cell*. Multiplying the rows by the columns will give us the number of cells in the grid (8x5=40 cells).

In order to work better, rows and columns are denoted respectively by the letters "f" and "c", to differentiate them from each other a subscript is used. So we write f1 to denote the first row (from top to bottom; fi, f_2, f_3, etc.), something similar we do with the columns, c1 indicates that it is the first column (from left to right; c_1, c_2, c_3, etc.). For cells we use a double subscript, the first one indicates the position of the row where the cell is and the second one the column. Thus c_{11} will indicate that the cell referred to is the one formed by the first row and the first column. Another example c_{23} is the cell formed by the second row and the third column. Illustration 2 shows these examples

Ilustration 2 Grid with its notation

	C_1	C_2	C_3	...	...
F_1	C_{11}	C_{12}	C_{13}	C_{14}	
F_2	C_{21}	C_{22}			
F_3					
:					
:					
:					
:					

Based on the elements described above, we are able to formulate a definition of a table. We will call a statistical table a **conveniently modified lattice set where the statistical analysis is performed**.

Parts of a Table

A table is divided into its external and internal elements. We will describe the minimum elements that compose both the internal and external elements. The external elements are elements that go outside the grid and the internal elements go inside the grid. See illustration 3

Ilustration 3 Partes de una tabla

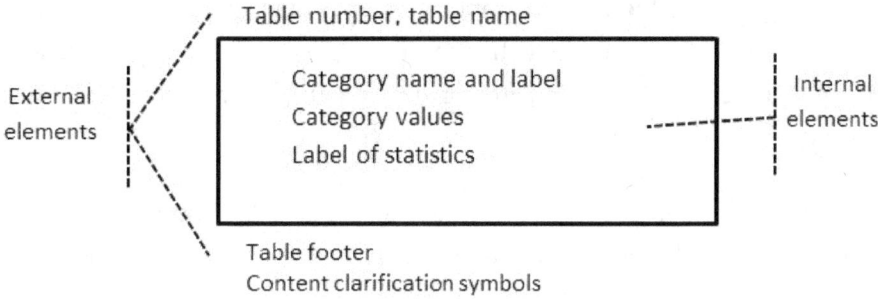

Figure1 Parts of a table according to constituent elements

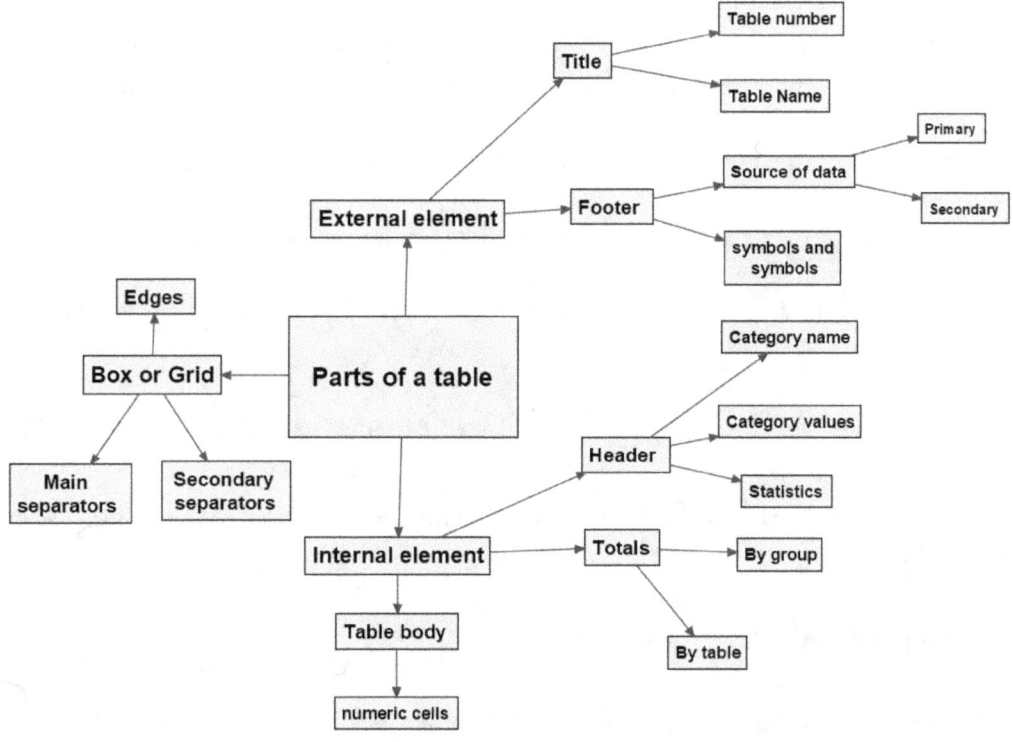

Table number

It is a numeral or letter that indicates the position of the table with respect to the other tables. This number can be an Arabic numeral (1, 2, 3, etc.), Roman numeral (I, II, III, IV, etc.) or simply Latin letters (a, b, c, etc.). It is usually followed by the word "Table.....", or by the word "table.....".

This number can be simple or compound. It is simple when there are not many tables (table 1, table 2, table a, table b, table I, table II, etc.) and it is compound when the document is long and divided into chapters, for this two numbers are placed (table 12, table Ia, etc.) where the first value indicates the chapter where the table is and the second the position with respect to the other tables. For example table 24 is in chapter 2 and of 4th positionIt is a numeral or letter that indicates the position of the table with respect to the other tables. This number can be an Arabic numeral (1, 2, 3, etc.), Roman numeral (I, II, III, IV, etc.) or simply Latin letters (a, b, c, etc.). It is usually followed by the word "Table.....", or by the word "table.....".

This number can be simple or compound. It is simple when there are not many tables (table 1, table 2, table a, table b, table I, table II, etc.) and it is compound when the document is long and divided into chapters, for this two numbers are placed (table 12, table Ia, etc.) where the first value indicates the chapter where the table is and the second the position with respect to the other tables. For example table 24 is in chapter 2 and in 4th position.

Table name

It is a description of the contents of the table. Normally institutions, universities or large companies have rules for naming a table. Here we will only show the most important elements of the name, i.e., to mention

1. Category labels displayed. E.g. Age range of the Ministry's employees.

2. The combination of the categories, if any. E.g. Age Range by Gender of the Ministry's workers.

3. The statistics used (which follow the category labels). E.g. Attitude of people towards change. Simple frequencies and percentages

4. The base on which the percentages are compared (if available): e.g. (Base: Total cases) or (Base: 345 interviews).

Veamos algunos casos de cómo se forma el título de la tabla dependiendo del número de variables que se exhiban y de su combinación

Case 1: The table shows only one variable.

The example below shows a table with the values of the family income of 185 families. As the table shows only one variable, the title would be: Family income of the workers of the "company Minas y asociados. Frequencies and percentages". Base: 185 workers

Table 11 Family income of workers of a group of workers (in Bs. F)

Less than 1000		From 1001 to 1500		From 1500 to 2000		Total	
N	%	N	%	N	%	N	%
45	24,32	65	35,14	75	40,54	185	100

Case 2: The table shows more than one variable.

The table below shows a table with two variables, gender and vehicle ownership. In this case the title would be: **Gender and Vehicle Ownership of the workers of the company Minas y Asociados**. Simple frequencies. Note that the base is not included as there are no percentages.

Another way to write the title is: Data of the workers of the company Minas y Asociados by Gender and Vehicle Ownership. Simple frequencies.

Table 12 Gender and Vehicle Ownership of 100 persons

Gender		Total Vehicle	Ownership		Total
Male	Female		Yes	No	
N	N	N	N	N	N
47	53	100	70	30	100

Case 3: Combined variables

The table below shows an example of a table with two variables in the horizontal part and one downwards (lateral part). In this case the nomenclature **POR-SEGÚN** that discriminates the horizontal and vertical variables is used. The name of the table in this technique would be: **Data of the workers of the company Minas y Asociados BY gender and age ACCORDING to the religion they profess. Frequencies and percentages (Base: total interviewees)**. In other words, the word BY is used to describe the variables that are placed at the top and the word BY is used to describe the variables that are placed at the side. One methodological reason is that the side variables are dependent and the horizontal variables are independent. Now the title can be abbreviated as **Gender and Age of the workers of the company Minas y Asociados SEGUN Religión que profesa. Frequencies and percentages (Base: total interviewees)**

SPSS uses a different nomenclature, the horizontal variables are described by the word THROUGH and the lateral variables by the word BY. In the example it would be **Data of the workers of the company Minas y Asociados BY gender and age THROUGH the religion they profess. Frequencies and percentages (Base: total interviewees)**

Table 13. Gender and age according to professed religion

| | | Gender | | | | Age | | | |
| | | Male | | Female | | More than 30 | | Less than 30 | |
		N	%	N	%	N	%	N	%
Religion	Catholic	58	53,21	60	48,39	63	57,27	26	45,61
	Evangelical	43	39,45	62	50	42	38,18	23	40,35
	Other	8	7,34	2	1,61	5	4,55	8	14,04
	Total	109	100	124	100	110	100	57	100

There are other more extensive ways of expressing the title that include the source of the data, for example, **Data of the workers of the company Minas y Asociados BY gender and age ACCORDING to the religion they profess, taken from the books of the Human Resources Department in August 2002**. This is unnecessary because it bulks up the title, the source should be in the table footer.

FOOTNOTES

Data source.

It refers to the origin of the data. The importance of the source is vital because a table is a document that can be audited to see the veracity of the data. A table without a certain source is useless because it does not communicate veracity..

Primary source.

It is the one where the data obtained has not yet been transformed by any statistical procedure, they are generally disseminated or published by the same researcher or institution that collects the data. Examples of primary sources are surveys, censuses and administrative records (medical records, accounting books, meteorological records, examination notes, etc.).

Secondary Source.

By opposition are all those data that are taken from publications, magazines, scientific journals, newspapers, reports, yearbooks, web sites, etc. Note that these are records in which the person seeking the information has not intervened to form the data, but which belong to other institutions or companies..

Notes, calls and conventional signs.

In order to make the table more user-friendly by clarifying certain aspects of the data shown or of the variables stated, symbols are generally used for this purpose, these can be asterisks (*), dashes (-), crosses (†), double cross (‡), letters (P, a), numbers (3), etc.... These symbols are divided into Notes, Callouts and Conventional Signs.

Notes

For the National Statistics Institute (INE), a note "is general information intended to provide concepts or definitions, clarify the content of tables or indicate the methodology adopted in the research or compilation of the data". It is indicated with the word NOTE:

Call.

It is information of a specific nature that clarifies some aspect of a number or variable text or variable values. The calls can be letters, numbers or symbols. Letters are used when clarifying some aspect of the numbers, numbers are used when clarifying aspects of the text, and symbols are used in both cases.

Other conventional signs and abbreviations.

A special type of callup are the conventional signs which serve to clarify special values in the cells. E.g. when blank cells or cells with zero value appear. Abbreviations are conventions used to summarize names of institutions, university degrees, etc. (e.g. Biology is summarized as bio; hectare, ha, etc.). Let's look at an example with all these table elements

The following example shown in Illustration 4 taken from the Institute of Statistics shows one way of using so-*called numbers* and so-*called letters* in a table. The Note clarifies that as of January 2014 (the table is much larger we only show the values for 2006) the projections will be adjusted to those of the 2011 Census. Hereafter it can be seen that the so-called numbers are next to the labels so that they are not confused with the letters and the *so-called 1* refers to figures for companies with 1 to 4 workers; similarly we have the so-called letters that appear next to the numbers so as not to confuse them with the numbers, the *so-called a* says that the marked figure refers to a coefficient of variation between 5 and 10 percent.

Ilustration 4.Scheme of call types in a table

Formal and informal sector, informal sector occupation category and gender	2006					
	January	February	March	April	May	June
TOTAL OCCUPIED	0	0	0	0	0	0
CLASSIFIABLE [Calling numbers]	-6.424	-610	-779	-2.019	-2.537	-3.624
(%)	-0,1	0,0	-0,1	0,0	-0,1	-0,1
FORMAL SECTOR	-4.925.644	-4.831.722	-5.027.789	-4.850.959	-4.822.965	-4.961.821
(%)	54,0	54,7	53,8	55,6	54,9	54,3
INFORMAL SECTOR 1/	4.919.220	4.831.112	5.027.010	4.848.940	4.820.428	4.958.197
(%)	46,0	45,3	46,2	44,4	45,1	45,7
NON-PROFESSIONAL SELF-EMPLOYED	3.066.428	3.027.436	3.117.299	2.942.741	3.012.413	3.109.211
(%)	28,6	28,4	28,7	27,0	28,2	28,6
EMPLOYERS	365.523 a/	373.637 a/	380.698 a/	363.352 a/	342.499 a/	435.817 a/
(%)	3,4 a/	3,5 a/	3,5 a/	3,3 a/	3,2 a/	4,0 a/
EMPLOYEES AND WORKERS 2/	1.307.274	1.306.210	1.378.065	1.403.744	1.350.461	1.326.895
UNCLASSIFIABLE 3/	6.424 c/	610 c/	779 c/	2.019 c/	2.537 c/	3.624 c/
(%)	0,1 c/	0,0 c/	0,0 c/	0,0 c/	0,1 c/	0,1 c/

SOURCE: Household Sample Survey; INE.

NOTE: As of January 2014 the total population is adjusted to the population projections according to the 2011 census.
1/ Includes people working in companies with 1 - 4 people. [Called letters]
2/ Includes domestic service that is not self-employed.
3/ They did not declare any of the variables that allow their classification in the Formal or Informal Sector of the economy.
4/ Differences in the variations of some rates or percentages are due to rounding of the figures considered in the calculation.
a/ This figure has a coefficient of variation greater than 5% and less than or equal to 10%.
b/ This figure has a coefficient of variation greater than 10% and less than or equal to 20%.
c/ This figure shows a coefficient of variation greater than 20%.
   The rest of the figures show variation coefficients of less than 5%.

Let's see other examples of conventional signs.

The value shown is insignificant. Table 1 A shows 9 cells with values. Suppose values 2 and 1 are insignificant, we could use a "Call" superscript letter to indicate this and put it in the table footer below the font, in this case the letter "a" was used.

The values displayed are less than a limit below which it is not worth displaying. In these cases it can be decided to set a Call symbol (dash) or to set the number zero. Table 1 (B1) shows the case when very small or insignificant values are displayed and a call symbol, dash (-) is put instead of the number. B2 shows the case when instead of putting a hyphen a zero is put. We must clarify that the decision about the limit or range in which the value is considered to be insignificant is arbitrary, it can be decided that the values that are not shown are those less than 10 or less than a decimal as 0.5 or 0.1 .... This technique is also used to "clean" the table of very small values when removing them highlights the larger ones.

Empty cells. Some statistical programs leave empty cells when they do not detect measurement (there was no count in that cell), this can be confusing because it can be interpreted as a lack of measurement in that cell (as the program does) or that the measurement was null, that is, that there is missing data or that the data is zero. The cell should be filled, for example, with a three-dashed call (...) (see Table 1C) and make it clear that there is missing data.

Preliminary (P), revised (R) or estimated (E) figures. There are cases in which the figures are not definitive because some places are not counted (as in the case of elections in which preliminary figures are given at a certain time of the night), another case is that the figures have been revised (case of tables that offer information and then another similar table appears with a different figure) or cases in which an estimate is made (indirect measurement); In all these cases, a so-called letter (capital letter) P should be placed next to the figure or in the heading of the table to indicate that it is a preliminary figure, R, to indicate that they are revised, and E for the estimated ones.

These aspects of notes, callouts and conventional abbreviations and symbols should not be lost sight of since, apart from clarifying the figures given, they give reliability to the table.

Box 1. Table composition. Three cases of table footer symbols table

A. Insignificant value expressed in letters

34	67	2[a]
67	345	97
81	1[b]	67

([a], [b]): Insignificant values

B1. Insignificant value expressed with dashes

34	67	-
67	345	97
81	-	67

(-):Insignificant values

B2. Insignificant value expressed with dashes

34	67	0
67	345	97
81	0	67

(0): figures less than 0.5

C. Empty Cells

34	...	34
67	345	97
81	78	...

(...):Information not available

Internal elements

TABLE HEADER

Code of the category or variable. We have already explained that this name may consist of a few letters to code the category. It is usually used together with digits indicating the position of the name in the database. E.g. Gen01, educ02, inces03, etc. For Gender01 it would be: variable name, gender and the 01 indicates that it is the first variable. This name or code is almost never written in the tables, it is only for database code (if it is written it is put before the category label).

Category label or variable name. This is the description of the variable and should be made up of two elements, the category itself and the unit where it is being observed. E.g. Gender of UNEFA students;

Educational level of Ministry workers. For the first case, "Gender of the students" is the name and "of the UNEFA" is the unit being observed (or where the data comes from). It is clear that the category is the subject of the sentence and the unit where it is measured, the complement of the sentence. In the table it appears as a heading for the category values (see Figure 3).

Category values. These are the variants of the category e.g. Gender of UNEFA students is the label of the category and "male" and "Female" are its values (its variation form); the age (label of the category) varies in years therefore the values are the years completed (12, 45, 78, etc.) and Educational Level varies in basic, medium, university, etc.

Labels of statistics or statisticians. It is the way they are denoted, e.g. simple frequencies are usually written as "N", "Count", "Count" or "Frequency". Percentage is written "%" or "percent". Mean as "χ" or "Mean", etc.

The set formed by the labels of the category, the values and the labels of the statistics will be called the **structure of the variable** or **variable** (what it is, how it varies and how it is measured).

TOTALS

Group total. When there are several categories in the table, the total of each of them is the group total. The importance of the totals lies in the fact that they verify the basis for calculating the statistics and their position gives guidance on the order of the sum. This will become clear later when we look at the tables and how they are constructed.

Own Totals. These are group totals of an item, variable or set of cells that we will later call *nest*. We will come back to this point

Total Table. This is the sum of the frequencies and percentages of the category variants. Totals should be placed next to the last variant or next to each category. This will be discussed later

Illustration 5.Internal table elements

	Student Gender [1]		
	Male[2]	Female[2]	Total
N[3]	Frequencies [4]	Frequencies[4]	
%[3]	Percentage's	Percentage's	

(1): Category Label or Variable Name

(2): Category or Item Values

(3): Statisticians or Statisticians Label

(4) Body of table

This is the set of cells containing the data to be analyzed. In Figure 6 it is represented by the four cells of frequencies and percentages.

Box or grid

The box or grid is always present as it is the space where the tabular analysis is performed (this can be seen in the grids of Excel or Word or any other program). The cells of the grid can be made visible by highlighting the lines or inserting convenient separators that highlight those aspects we want. This gives rise to the separators and borders that make up the styles of the table.

MAIN SEPARATORS

These are the horizontal or vertical lines that divide the header and totals from the body of the table.

SECONDARY SEPARATORS

These are the horizontal or vertical lines that divide the header or totals internally and may or may not cross the body of the table or data.

BORDERS

They are the lines that separate the table from the rest of the context of the sheet, they can be top, bottom, left or right border.

The following table shows what has been described, the double lines indicate the main separators, the thick lines the borders and the simple lines the secondary separators.

Box 2.Lines used in a table

		Judge			
		Good	Medium	Bad	Total
Juez 1	Good	18	11	6	35
	Medium	11	30	9	50
	Bad	2	3	20	25
	Total	31	44	35	110

⸻	Simple line:	Secondary separators
▬	Thick line:	Edges
═	Double line:	Main separator

Chapter 4. Tabular Logic

Tabular definitions

Some of these definitions were already given in chapter 3 when defining the parts of a table, especially the Internal Elements of a table that have to do with the formation of the Complete Header, so called because some tables do not have all of these elements, for example, some do not have the totals.

Full Headline

It is the structure that has.

1. Variable name (A). pe., Gender

2. Category or item scale values (ai, bi). pe., male, Female

3. Statistics labels; denoted in general form as "e", pe. pe. N, %.

4. Labels of the totals (t). there are different types of totals, which we will see in due course.

A complete heading has the structure $<A, i, e, t>$ in the following order: first, A, in front of this goes aii, next to these goes t and in front of these goes e. Some institutions place the totals before the items.

Frequency Table

When a complete heading, A, has data cells in front of e and t, we will call it a *frequency table* and we will use the same notation as for the complete heading (A, B, C, ...), since we assume the existence of data when speaking of heading.

Holistic tabular structure.

It is a structure formed by several tables, A, B, C, etc.; we will denote it using the Greek capital letters Ψ, Σ, Ω, etc. They are each designed and configured in such a way that some are contained in others, intercepting, combining or going together; when this happens the structure is configured in holons ranging from an item (i) to a total (t). Box 3 shows several holons of the structure $A \mid (B/C)$ these are:

Boc 3. Visualization of holons in a tabular structure $\Sigma =_2 A \mid (_2 B /_2 C)$

		B								
		b_1			b_2			Total		
		C			C			C		
		c_1	c_2	Total	c_1	c_2	Total	c_1	c_2	Total
		N	N	N				N	N	N
A	a_1			a_1b_1						
	a_2									
	Total	c_1b_1			c_2b_2					

1. $a_1 \rightarrow a_1b_1$
2. $a_2 \rightarrow a_2b_1$
3. $c_1 \rightarrow c_1b_1$

These tables A, B crossed and intercepted form internal structures that we call *nests* formed by the frequencies and *layers* formed by the names of the variables. Since 'tables', are really structured sets (of tables), and since they are generically known by that name, we will call them indistinctly tables or tabular structures unless we need to clarify the terms

Layer and nesting

These two terms are used in computer science as programming commands. In statistical tables they possibly have a similar sense, however, we will use them methodologically as processes of construction of the tables because they bring confusion to the automatic transfer of programming. We will resume the concepts in the part of nested tables, but here we will make a definition in methodological terms.

A layer is a composition of functions that is mathematically written as

$F(G(x))$, for example, if $G(x) = x + 3$ y $F(y) = y + 5$,

Then $F(G(x)) = G(x) + 5 = x + 3 + 5 = x + 8$.

De los dos elementos involucrados en la relación de composición las Capas son las funciones F y G, las reglas de correspondencia $x + 3$ e $y + 5$ son los elementos anidados siendo $x + 8$ el resultado del anidamiento. En el caso de las tablas estadísticas, las reglas de correspondencia son las frecuencias dobles, triples, ... n-ples o $f_2, f_3, ... f_n$ y las Capas las funciones F y G, el resultado $(x + 8)$ es en tablas el Total de la Capa. De esta manera tenemos tres definiciones involucradas en la dinámica tabular, la capa, el nido y las frecuencias, igualmente tendrá relación con los encabezados completos pues es estructurante. Sin embargo, para dar coherencia a los nombres de los totales evitando un caos nominal omitiremos la mención de totales con nombres de fila, columna, capa, entre otros, en su lugar estandarizaremos los nombres como *totales de nidos* pues la estructura tabular con el cruce de tablas forma anidamientos incluso en las tablas simples llamaremos nido a las celdas que contienen valores y su total un total de nido.

Of the two elements involved in the composition relation the Layers are the functions F and G, the correspondence rules x+3 and y+5 are the nested elements with x+8 being the result of the nesting. In the case of statistical tables, the correspondence rules are the double, triple, ... n-frequencies or f2, f3, ... fn and the Layers are the F and G functions, the result (x+8) is in tables the Layer Total. In this way we have three definitions involved in the tabular dynamics, the layer, the nest and the frequencies, it will also be related to the complete headings because it is structuring[5]. However, to give coherence to the names of the totals avoiding a nominal chaos we will omit the mention of totals with names of row, column, layer, among others, instead we will standardize the names as *nest totals* because the tabular structure with the crossing of tables forms nestings even in the simple tables we will call nest to the cells that contain values and its total a nest total.

Tabular design

By *table design* we mean the arrangement of variables in the tabular space according to the required analysis. This involves two aspects.

1. The arrangement of the variables in the tabular space according to their relational complexity, whether they are stacked, nested or two-dimensional, and 2.
2. The spatial arrangement that allows optimizing the sheet where the table is going to be read.[6]

Ilustration 6. Classification according to the spatial design of the table. Spatial classification SL

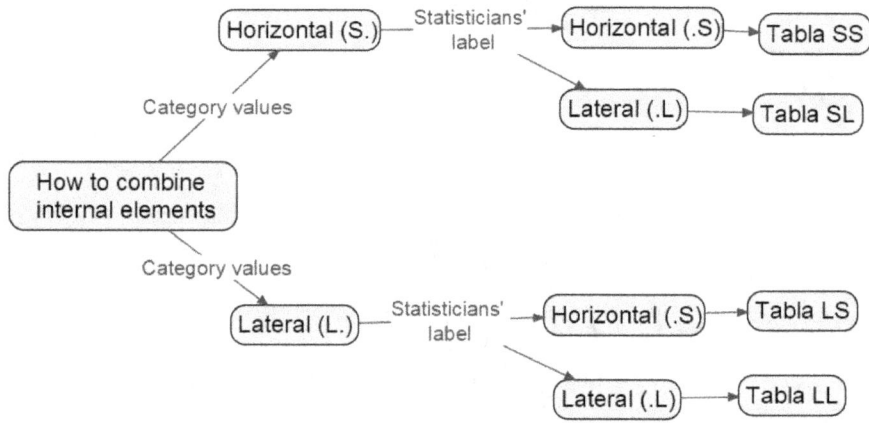

A table is pre-designed by placing, conveniently in the matrix, the values of the category and the label of the statistics, once this is done the table is practically designed. These elements (values and labels) can only be placed in the matrix in two ways, horizontally and vertically.

When placed horizontally we say that they are placed at the top of the table (S) and when placed vertically we say that they are placed at the side (L) of the table. In the same way, the labels of the statistics are placed in both positions S or L. Thus we have four table designs, those with the category values and statistic labels in S position are denoted SS tables; those with the category values in S position and statistic labels in L position are denoted SL

---

[5] It is important to note that in some statistical software manuals some totals are called layer, row layer, among others, which are confusing because there is no explicit definition.
[6] This research was done by looking at the tabular arrangements of some statistical programs to achieve agreement, but the theoretical approach of these programs is very poor.

tables. In the same way we have the LL and LS configurations which are nothing else than the transposed tables of the previous ones.

Tables 4 and 5 show these positions. Note that the S or L positions of the category values are relative to the body of the table or data cells, which is obvious since these values have to "look" at the data. In example 22 A1 the category label is above the values and in example 22 A2 it is on the side, but the category values are always behind the body of the data. Similarly for examples B1 and B2, in the first, the label (Gender) is on its side and in the second it is above its values and again the category values are behind the body of the table. In all examples A1, A2, B, and B2 the position of the category values (S or L) is relative to the body of the table.

Box 4. Top and side position of the variable labels

A1. Superior (S)

Género		
Male	Female	Total

A2. Superior (S)

	Male	Female	Total
Gender			

B1. Lateral (L)

	Male	
Gender	Female	
	Total	

B2. Lateral (L)

	Gender
Male	
Female	
Total	

With respect to the totals, these should be placed after the values, this is automatic, write the values in order and then the total, Male → Female → Total. Examples 4 and 5 (A1, A2, B1 and B2) show this procedure, when writing the values the total is placed immediately in the adjacent column or row. This procedure is important because the position of the total indicates the path of the sum of the values of the labels or scales used. This is particularly important when the tables are large and collect many variables.

Now we only need to place the labels of the statistics which go in front of the category values and in position S or L with respect to them. Let's see example 18, the position S and L of the labels of the statistics of table A1 and A2. Note also that we have used as labels for the statistics, frequency (N) and percentage (%).

Box 5. Top and side position of the labels of the statistics in Table A

A1 (SS)

Gender					
Male		Female		Total	
N	%	N	%	N	%

A2 (LS)

		N	%
	Male		
Gender	Female		
	Total		

For the SS configuration, the labels of the statistics are placed horizontally below the values, while for the LS configuration the labels of the statistics go in front of the values, but one row higher. This is in order to cover all rows. In the SS case it would not be possible to cover all the columns. It remains for the reader to complete the other two tables, LL and SL (the transposes of SS and LS).

Tabular configuration.

It is a way of combining different headings A, B, C, etc., for the achievement of the objectives. Tables can be configured in two ways: stacking two or more headings and nesting them; nesting can occupy both dimensions of space as we shall see; nesting can also be stacked and nested. Two tabular structures are similar if they have the same configuration; this implies that their change of shape does not alter the internal relationships of the headings and percentages; i.e. they remain the same.

Header stacking

Two tables are stacked when their headers are stacked; that is, when they are put together in S or L position. To indicate that two headers are stacked we will use the asterisk symbol "*"; A*B*C although we can omit it by understanding it in the ABC relation; the stacking can be recursive and we will denote (A*B)s when it occurs horizontally and l when it is transposed.

One-dimensional nesting

Two tables are nested when the items of a nesting variable contain the items of the nested one. Table 2 shows an example (B/C) (part of table A|(B/C)). We will use the slash symbol "/" to denote that one variable A nests another variable B, thus, A/B, and read "A over B". Likewise, the nesting is recursive, A/B/C/.../. The nesting variable is also called layer; in the example; however, as there is a recurrence of variables in layers we will use the terms 1st, 2nd layer, etc.; the structure A/B/C, has a 1st layer B and a second layer A or major layer.

Two-dimensional nesting or cross-tabulation

Double-entry tables are doubly nested by tabular holism. We will use in sidebar symbol "|" to denote the double nesting, thus, A|B and read "A vs. B"; the heading A is in position L, while B, in position S, see Table 15; therefore, we will only use the letters S or L to indicate the position of e. The example shows two-dimensional nesting; note the holons contained with each other and all under $\sum$; for example (b1→113; a1 →105; (A, B) →215). The two-dimensional nesting is not recursive because of spatial impedance..

Table 14. Basic two-dimensional chart model $\sum = {}_2A \mid {}_2B$

		B		
		b_1	b_2	Total
		N	N	N
A	a_1	53	52	105
	a_2	60	50	110
	Total	113	102	215

Tabular formulae

A tabular formula is a series of symbols that describe the configuration of the structure using 1. the complete headings A, B, ...; 2. the structure connectors (*, |, /), 3. the symbols of tabular design (S, L) and 4. the number of items or values of the variables (a1, a2, a3...) that will be a value or cardinal of the number of items or values of the variables (a1, a2, a3...) that will be a value or cardinal of the number of items (a1, a2, a3...).) which will be a value or cardinal of the number of items, so for example in the structures, (2A*3B)ss, (3A/2B)ll and (2A/3B)|3C)sl, it is indicated for the first one that the variables A and B are stacked with 2 and 3 items in position S and the statistics also in position S; in the second example we have two nested variables A and B, the first variable with 2 items, the second with 3 in position S and the statistics in position L, the reader should do the third one.

Tabular dynamics

A Σ retains its overall total and its headings in all its possible configurations; holons can move in the tabular space resignifying Σ according to the objectives of the study and producing families of structures ($\Sigma1$, $\Sigma2$, $\Sigma1'$, $\Sigma2'$, $\Sigma2'$, ...). These motions are, rotation, translation, segmentation and distribution.

ROTATION

The first, *rotation*, consists of moving the holons from an S to an L position (see tables 15 and 16), which implies moving from an A|B to a rotated B|A structure.

Table 15. Structure $\Sigma = (_2A|_2B)$

		B		
		b_1	b_2	Total
		N	N	N
	a_1	53	60	113
A	a_2	52	50	102
	Total	105	110	215

Tabla 16. Rotated structure $\Sigma^r= (_2B|_2A)$

		A		
		a1	a2	Total
		N	N	N
	b1	53	52	105
B	b2	60	50	110
	Total	113	102	215

TRASLATION

Translation involves moving the holons from one S or L position to another, without rotating it; that is, moving them from a stacked (L) to an aligned (S) position, holon by holon or by group of holons which involves transforming the double-entry A|B structure to a nested A/B one by breaking the double nesting by one of them. Table 17 shows the Σ structure of graph 1 transformed.

Table 17. Structure $\Sigma^t =(_2A/_2B)s$

A								
a_1			a_2			Total		
B			B			B		
b_1	b_2	Total	b_1	b_2	Total	b_1	b_2	Total
N	N	N	N	N	N	N	N	N
53	60	113	52	50	102	105	110	215

The $\sum$ *segmentation* consists of splitting the structure when it is very heavy. Segmentation occurs in two ways, by *splitting* the nesting layer or variable into its items and by *partitioning* two stacked tables, Table 18 shows the segmentation by splitting and Table 19 by partitioning

Table 18. Segmentation by tabular division of the H = $({}_2A/{}_2B)$ en $\sum a_1$, $\sum a_2$ y $\sum T$

	A								
	a_1			a_2			Total		
	B			B			B		
	b_1	b_2	Total	b_1	b_2	Total	b_1	b_2	Total
N	53	60	113	52	50	102	105	110	215

H= $a_1/{}_2B$			
	B		
	b_1	b_2	Total
N	53	60	113

H = $a_2/{}_2B$			
	B		
	b_1	b_2	Total
N	52	50	102

$\sum$= Total			
	B		
	b1	b2	Total
N	105	110	215

Table 19.Segmentation of X= $_2A\,|\,({}_2B_2C)$ en X´= $({}_2A\,|\,{}_2B)$ y X´´= $({}_2A\,|\,{}_2C)$ by partition

			B			C		
			b_1	b_2	Total	c_1	c_2	Total
	a_1	N	53	60	113	64	49	113
A	a_2	N	52	50	102	65	37	102
	Total	N	105	110	215	129	86	215

X´= $({}_2A\,|\,{}_2B)$

			B		
			b1	b2	Total
	a1	N	53	60	113
		%	46,9	53,1	100
A	a2	N	52	50	102
		%	51	49	100
	Total	N	105	110	215

$X'' = (_2A|_2C)$

			C		
			c_1	c_2	Total
A	a_1	N	64	49	113
		%	56,6	43,4	100
	a_2	N	65	37	102
		%	63,7	36,3	100
	Total	N	129	86	215

DISTRIBUTION

Finally, the *distribution* is the reassignment of items from one holon to another when a nested variable becomes a nesting variable; that is, reassigning them from the first level to a layer or vice versa (see Table 20), the distribution implies the disintegration of the nests into others and the reconfiguration of the totals as the new layer calculates its own totals.

Table 20.Distribution of the items of the form $(_2A/_2B)$ en $(_2B/_2A)$

$_2A/_2B$	A								
	a_1			a_2			Total		
	B			B			B		
	b_1	b_2	Total	b_1	b_2	Total	b_1	b_2	Total
N	53	60	113	52	50	102	105	110	215

$_2B/_2A$	B								
	b_1			b_2			Total		
	A			A			A		
	a_1	a_2	Total	a_1	a_2	Total	a_1	a_2	Total
N	53	52	105	60	50	110	113	102	215

Tabular Dynamics Characteristics

Five characteristics are fundamental in Tabular Dynamics,

1. The conservation of the grand total of the tables
2. The commutativity of the terms or semantics, that is, *(abcd) = (bacd) = (cabd)=......*
3. This allows the tabular structure formed relations between tables $\Sigma = (A, B, C,...)$ to conserve the quantities per cell regardless of the configuration of the structure
4. While the cells retain semantic quantities the *translation, rotation* and *segmentation* features make the structure a methodological tool for descriptive data planning and analysis
5. The *distribution* feature adds a change of focus to the analysis as it redistributes the totals simplifying the analysis when the structure is complex. That is, the first three moves change the topology (shape) of the structure and

the last one changes the configuration. The tabular dynamics has objectives of optimization of the tabular space and, methodological to simplify complex structures.

Tabular Dynamics of some structures [7]

$$\text{ESTRUCTURA } M = (_3A/_2B)$$

Los subíndices 3 y 2 indican que las variables A y B tienen respectivamente 3 y 2 semánticos (a_1, a_2, a_3 y b_1, b_2) por lo tanto un total por cada trio y dupla de ítems; como la estructura está anidada (/) las celdas están formadas por las frecuencias dobles, f_2, de los ítems (a_1b_1, a_1b_2, a_2b_1, a_2b_2, a_3b_1, a_3b_2) que son combinaciones ordenadas. Estas f_2 se conservan cuando la estructura cambie su forma, esto es, la forma A/B está en posición S o L, $(_3A/_2B)$s o ($_3A/_2B$)l desplegadas en una sola dimensión del espacio, sin embargo puede hacerlo en las dos dimensiones del espacio, $(_3A|_2B)$ o $(_2B|_3A)$ sin alterar sus frecuencias dobles f_2 pero si reacomodándolas en la estructura y por lo tanto los totales que en la primera forma son Ta_1, $(a_1b_1+a_1b_2)$, Ta_2 y Ta_3 luego los de Tb_1 y Tb_2 y el Tg para 6 totales, en la segunda son Tb_1, $(a_1b_1+a_1b_2+a_1b_3)$, y Tb_2, adicionalmente los Ta_1, Ta_2, Ta_3 y Tg son igualmente 6.

Este reacomodo de las f_2 y de los totales tiene implicaciones metodológicas. La tabla 21 muestra la dinámica tabular de las estructuras formada por $_3A$ y $_2B$, las primeras dos formas son de distribución y la tercera de traslación, la segmentación sería por división, si dividimos la estructura mediante B, sería, $(b_1/_3A)$ ó $(b_2/_3A)$ y Tb (a_1, a_2, a_3, tg), el lector puede hacer la división por A y la de A | B

Tabla 21. Formas $(_2B/_3A)$l, $(_3A/_2B)$l y $_2B|_3A$

1. $_3A/_2B$

A				
	a_1	B	a_1b_1	39
			a_1b_2	31
			Total	70
	a_2	B	a_2b_1	42
			a_2b_2	38
			Total	80
	a_3	B	a_3b_1	32
			a_3b_2	33
			Total	65
	Total	B	b_1	113
			b_2	102
			Total	215

2. $_2B/_3A$

B				
B	b_1	A	b_1a_1	39
			b_1a_2	42
			b_1a_3	32
			Total	113
	b_2	A	b_2a_1	31
			b_2a2	38
			b_2a_3	33
			Total	102
	Total	A	a_1	70
			a_2	80
			a_3	65
			Total	215

4. $_2B\,|\,_3A$

		A			
		a_1	a_2	a_3	Total
		N	N	N	N
B	b_1	39	42	32	113
	b_2	31	38	33	102
	Total	70	80	65	215

$$\text{STRUCTURE } \Omega = {}_3A\,/\,({}_2B *{}_4C)$$

The form $3A/(_2B*_4C)$, in this case we will omit the graphs. This structure consists of a nesting (/) and a stacking (*). The first cell has the semantic stacking $(a_1b_1, a_1b_2, ta_1)(a_1c_1, a_1c_2, a_1c_3, a_1c_4, ta1)$ and the amount of totals are 3 of A for each 2 of B plus the 3 of B , would be 6 (3+3) of A/B, on the other hand 3 of A for each 4 of C plus 5 of C would be 8 (3+5) of A/C in total are 8+6 = 14 totals.

As there is a stacking, the semantics of B and C are not combined, therefore they can be segmented by separation in $(_3A/_2B)$ and $(_3A/_4C)$ maintaining the amount of totals.

Holistic table taxonomy

Family

Let the headings or tables A, B, C, ..., and their items, i, j, k, we define *family tabular* to the set $_iA_jB_kC...$; where the subscripts i, j, k, ... represent the number of items or semantics of each heading or table pe., $_2A_2B_2C$, $_3A_2B_2C_4D$. in these examples we have respectively three and four conjugate tables in the structures $\Sigma = {}_2A_2B_2C$ and $\Psi = {}_3A_2B_2C_4D$.

Class

We define *tabular class* as the set of all permutations of a family; i.e., the above family has 3! = 6 classes (see Box 6). Each class has several configurations of subclasses.

Group

Given a Subclass of holistic structures or tables, the *group* is defined by the possible designs of each element of the Class. Box 6 shows the Family of tabular structures for three tables A, B and C..

Box 6. Family $_2A_2B_2C_2$ with its classes and tabular groups. For simplicity, the items in the subclasses are not shown.

Class	Subclass			Group
$_2C_2B_2A$	C|(B/A)	(C/B)|A	C/B/A	SS, SL, LS, LL
$_2B_2C_2A$	B|(C/A)	(B/C)|A	B/C/A	SS, SL, LS, LL
$_2B_2A_2C$	B|(A/C)	(BA)|C	B/A/C	SS, SL, LS, LL
$_2C_2A_2B$	C|(A/B)	(C/A)|B	C/A/B	SS, SL, LS, LL
$_2A_2C_2B$	A|(CB)	(A/C)|B	A/C/B	SS, SL, LS, LL
$_2A_2B_2C$	A|(BC)	(A/B)|C	A/B/C	SS, SL, LS, LL

The importance of the classes and subclasses lies in the fact that they offer different analyses because, having the same configuration, the totals against which the values are relativized change. This point will be discussed further in the part on semantic construction. Box 7 shows the basic structures

Boc 7.Basic formulas and their meaning

formula	meaning
(A*B) (x x)	Stacking
(A/B) (x x)	One-dimensional nesting
(A|B) (x)	Two-dimensional nesting

From these basic structures, more complex combinations or configurations are made.

Tabular validation

A *formula* is valid when it does not contradict the tabular structure created and is invalid when it contradicts it or when it is not possible to construct the referent table. The basic formulas in Table 7 are valid; equally valid are those that proceed by substitution of the headings by more complex structures. Some examples of invalid formulas are: A|B|C; it is invalid since the sidebar | admits only one antecedent variable in position L and one successor in position S, A|B; since both occupy the real two-dimensional space. Another example is, (A|B)*C; it is invalid because C cannot be constructed stacked to the two-dimensional sharing elements of the structure; the valid expression is A|(B*C) or (A|B)*(A|C).

Tabular equivalency

Two formulas are *equivalent* if the same tabular structure can be expressed by two different formulas. For example, A|(B*C) ≡ (A|B)*(A|C); Table 22 shows the structure.

Table 22. Tabular structure A | (B*C) or (A | B)*(A | C)

			B			C		
			b_1	b_2	Total	c_1	c_2	Total
A	a_1	N	53	60	113	64	49	113
	a_2	N	52	50	102	65	37	102
	Total	N	105	110	215	129	86	215

Another example is, (A*B) | (C*D) ≡ (A | (C*D))*(B | (C*D) ≡ (A | C)*(A | D)*(B | C)*(B | C)*(B | D). Table 23 shows this structure

Table 23. Tabular structure (A*B) | (C*D)

			C			D		
			c_1	c_2	Total	d_1	d_2	Total
A	a_1	N	64	49	113	58	55	113
	a_2	N	65	37	102	54	48	102
	Total	N	129	86	215	112	103	215
B	b_1	N	63	42	105	54	51	105
	b_2	N	66	44	110	58	52	110
	Total	N	129	86	215	112	103	215

Table Types

For this classification we will use as a starting criterion the use of space by the categories, that is, how the categories fill the space of the sheet or screen that physically is two-dimensional, however for the case of tables the dimensionality refers to the number of variables that are combined so that frequency tables $f_1, f_2, f_3,...$, correspond to tables of 1, 2, 3 or more dimensions then the way the categories are combined in the table. That is, in one dimension one or more variables can be stacked or nested. In two dimensions (rows and columns) stacked and nested arrangements can be made, which requires more information and more complex analysis, in fact, multivariate analysis can be used. Finally, in three dimensions the same stacked and nested arrays can be made, these tables are much more complex. See Figure 2

Figure 2Types of table according to the way the categories are combined

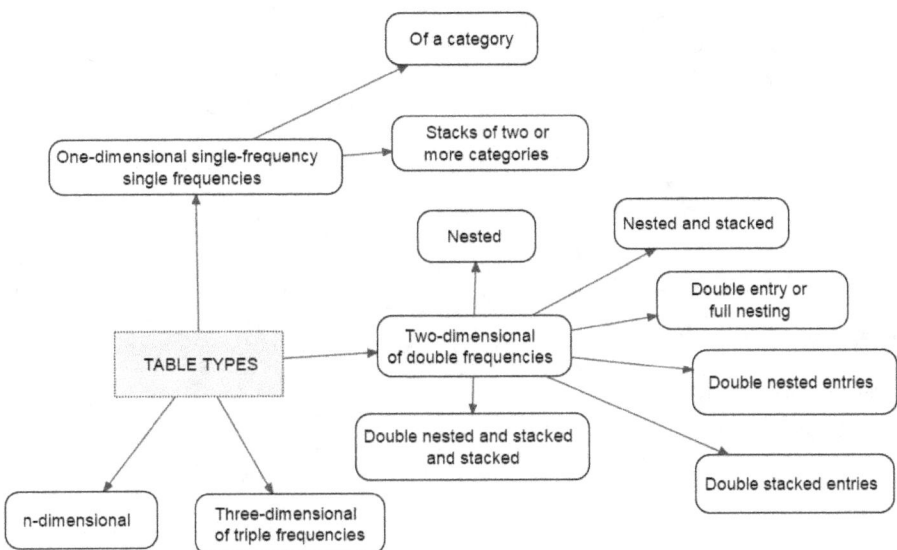

One-dimensional

One Category.

One-category tables, as the name implies, are small tables that show a single category. E.g. age of students, family income of workers, etc. The group total is the same as the table total. Example 14A shows a one-category table with the values of the category, the statistic (N) and the data. The heading occupies the first three lines and the last one, the body of the table.

Stacked.

The stacks show various categories side by side forming a stack, e.g. gender, age, income, etc. Each category has group totals and table totals if necessary. Let's look at an example of each of the tables described. Table 24 shows an example of a simple stacked table of two categories. Note the presence of totals per group; the heading occupies the first three lines and the body of the data, the last one. In practice we will understand that the simple ones are of one category and the stacked ones of two or more (both are simple because they do not combine the variables).

Table 24.Examples of tables. A of one category. B stacked of two categories. Hypothetical data

A. Table of a category

Family income of workers (in Bs. F.))			
Less than 1000	From 1001 to 1500	From 1500 to 2000	Total
N	N	N	N
45	65	75	185

B. Tabla Stacked two categories

Gender	Total	Vehicle ownership	Total

Male	Female		Si	No	
N	N	N	N	N	N
47	53	100	70	30	100

Two-dimensional

Nested

Seeing above the definition of nesting and its relationship with the layer, we will clarify these concepts further. We call them nested tables nominally by tradition and ease of expression, but they involve nesting, layers and frequencies. A *nesting* consists of placing within the values (or items) of the category of one variable (the one that nests) those of another (that is nested). See Figure 3

Figure 3 .Value nesting scheme

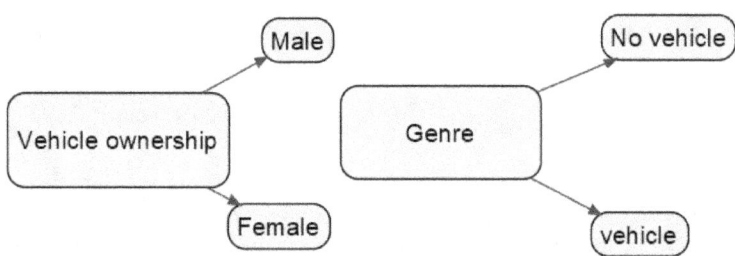

Figure 3 shows how the *Gender category* nests the *Vehicle Ownership* category and how the *Vehicle Ownership* category nests the *Gender category*. In the first case nesting Vehicle Ownership in Gender means counting the double frequencies "Female" and "Has Vehicle", likewise counting the double frequencies "Female" with "Does Not Have Vehicle". Likewise, for male-has vehicle and male-does not have vehicle. In the second case, nesting Gender in Vehicle Ownership means counting the pairs of values "Yes Has Vehicle-Male" on the one hand, and on the other hand the pairs "No Vehicle-Female" and so on.

Let's look at table 25, the structure of the table is $(_2G/_2T)s$ where variable Gender is Housing Tenure layer, the conjugate values in f2 of the nests are Male-have = 64, Male-have-not = 49 and same for the next nest, Female-have = 65 and Female-have-not = 37.

Table 25. Gender nesting on housing tenure, $_2G/_2T$

Genre								
Male			Female			Total		
Vehicle Ownership			Vehicle Ownership			Vehicle Ownership		
Has	Has Not	Total	Has	Has Not	Total	Has	Has Not	Total
N	N	N	N	N	N	N	N	N
64	49	113	65	37	102	129	86	215

In the following table 26 the nesting is inverted to T/G, observing the recalculation of the totals by reconfiguration of the nests or groups keeping constant the double frequencies or f_2. Evidently, the configuration of the table changes because the totals are not the same although the grand total is maintained.

Tabla 26. Gender nesting on housing tenure, $_2$T/$_2$G

Vehicle Ownership								
Has			Has Not			Total		
Genre			Genre			Genre		
Male	Female	Total	Male	Female	Total	Male	Female	Total
N	N	N	N	N	N	N	N	N
64	65	129	49	37	86	113	102	215

Double Entry.

It is also called a cross-tabulation table, but the complexity it acquires makes this name obsolete. A table of this type is read from the category labels to the numerical values (frequencies or percentages). Double-entry tables can be read from the S position, by columns, as well as from the L position, by rows. The nesting of tables 25 and 26 are partial (Gender in Housing Tenure or Housing Tenure in Gender). In the case of a double entry table the nesting is complete. This can be seen when obtaining the percentages; in the nested tables only the percentage of the nesting can be obtained (over the total of the nesting category), but in the double entry table three percentages can be obtained in the classic analysis, that is why the nesting is said to be complete, although placing two or more percentages causes confusion.

The reader can check each nest and verify that the f_2 frequencies are maintained in the double-entry table.

Table 27. Double-entry or contingency table $_2$G | $_2$T

			Vehicle Ownership		
			Has	Has Note	Total
Genre	Male	N	64	49	113
	Female	N	65	37	102
	Total	N	129	86	215

Source: Data processed from Visauta Vinacua (1997) CD database Statistical Analysis with SPSS for Windows.... File work.sav

Double-entry table of stacked categories

This double-entry table with stacked categories is quite common. There is usually a need to compare demographic categories (sex, age, country, etc.) against some psychographic, attitudinal, etc. category or variable to see how it varies with respect to the demographic. In this case, Table 28 shows two stacked demographic variables $(_4A\,|\,_3B)(_4A\,|\,_2C) = A_1\,|\,(_3B_2C)$, as they are stacked each is of frequency f_2. Note that we have placed the totals by group for the three variables, the totals will not be similar because variable B: Religion has missing values that do not combine with those of C: Sex.

Table 28. Double entry table with stacked categories. $A_1 \mid (_3B_2C)$

			B. Religión			Total	C. Genre		Total
			Catholic	Other	None		Male	Female	
			N	N	N	N	N	N	N
A_1: Concerns: aspect that worries you the most	Your emotional life		53	2	8	63	38	25	63
	Money		248	10	28	286	174	116	290
	Family harmony		255	4	18	277	113	167	280
	Your health		476	8	34	518	235	290	525
Total de group			1032	24	88	1144	560	598	1158

Source: Data processed from Visauta Vinacua (1997) CD database Statistical Analysis with SPSS for Windows.... File work.sav

A further degree of complexity is given by the stacking of four variables in a two-dimensional one. Table 29 shows this stacking in a *lego game* effect

Table 29. Stacking of 4 variables in a structure $\Sigma = [(_4A_1 \mid _3B)(_4A_1 \mid _2C)] \mid [(_4A_2 \mid _3B)(_4A_2 \mid _2C)]$

			B. Religión				C. Genre		
			Catholic	Other	None	Total	Male	Female	Total
A1 Concerns: aspect that worries you the most	Your emotional life	N	53	2	8	63	38	25	63
	Money	N	248	10	28	286	174	116	290
	Family harmony	N	255	4	18	277	113	167	280
	Your health	N	476	8	34	518	235	290	525
	Total	N	1032	24	88	1144	560	598	1158
A2 Concerns: aspect that worries you the least	Your emotional life	N	185	6	20	211	105	109	214
	Money	N	369	6	23	398	170	231	401
	Family harmony	N	148	6	8	162	84	80	164
	Your health	N	143	6	22	171	100	72	172
	Total	N	845	24	73	942	459	492	951

This structure Σ can be built thanks to the fact that the nests have internal totals that allow the independence of each structure while preserving the double frequencies f_2. A table with this structure can be segmented by structure partitioning.

Three-dimensional

Three-dimensional ones are formed with triple frequencies f_3 can be stacked and nested, with full nesting and stacking. Let's see some cases

Double entry table with nested categories.

This table adds a level of complexity by moving from structure $_4A\,|\,(_3B_2C)$ to 11 $(_4A\,|\,(_2B/_3C)$ of the same family. Table 30 shows the nesting sex on religion against the variable "worries: aspect that worries you the most". Note that the aggregate complexity implies frequencies f_3. Thus, the extent decreases by increasing the intensity, note the reader the nesting figures

Table 30. Double-entry table with nested structure categories $(_4A\,|\,(_2B/_3C)$

		C: Genre							
		c_1: Male				c_2:Female			
		B: Religión				B. Religión			
		Catholic	Other	None	Total	Catholic	Other	None	Total
A1 Concerns: aspect that worries you the most	Your emotional life	31	2	5	38	22		3	25
	Money	152	5	14	171	96	5	14	115
	Family harmony	103	1	7	111	152	3	11	166
	Your health	203	2	27	232	273	6	7	286
Total group		489	10	53	552	543	14	35	592

Source: Data processed from Visauta Vinacua (1997) CD database Statistical Analysis with SPSS for Windows.... File work.sav

Double-entry table with nested and stacked categories.

For this type of table we add another level of complexity, to the variable "worries: aspect that worries you the *most*" we stack the variable "worries: aspect that worries you the *least*", that is now we have the structure $(_4A_1\,|\,(_2C/_3B)(_4A_2)\,|\,(_2C/_3B)$. this is due to the conservation of the totals per nest. See example in table 31

Tabla 31. Double-entry table with nested and stacked categories

		C: Genre							
		c_1: Male				c_2: Female			
		B: Religión				B: Religión			
		Catholic	Other	None	Total	Catholic	Other	None	Total
A1.Concerns: aspect that worries you the most	Su vida afectiva	31	2	5	38	22		3	25
	El dinero	152	5	14	171	96	5	14	115
	Armonía familiar	103	1	7	111	152	3	11	166
	Su salud	203	2	27	232	273	6	7	286
	Total	489	10	53	552	543	14	35	592
A2.Concerns: aspect that worries you the least	Su vida afectiva	85	2	16	103	100	4	4	108
	El dinero	151	3	15	169	218	3	8	229
	Armonía familiar	73	3	6	82	75	3	2	80
	Su salud	88	2	9	99	55	4	13	72
	Total	397	10	46	453	448	14	27	489

Source: Data processed from Visauta Vinacua (1997) CD database Statistical Analysis with SPSS for Windows.... File work.sav

Two-dimensional table nested by items of a third variable.

Note in table 32 that in appearance the table is a double entry table "Religion" vs. "Concerns: aspect that worries you the most", but this two-dimensional table is nested by the item b1 Female of the sex variable, some statistical programs make them and refer to them as layers, but this is confusing, what happens according to the tabular logic is a segmentation of table 13 by division of the items. Note the label of the layer (Sex b_1 Female) in the upper left corner of the table Note that this table 13 is the same table of example 11 Female-religion segmented (see data and totals), we will see this further on

Table 32. A. Contingency table with category in structure layer $b_1/(_4A\,|_3B)$

Genre b_1 Female					
				B: Religión	
		Catholic	Other	None	Total
A1: Concerns: aspect that worries you the most	Your emotional life	22		3	25
	Money	96	5	14	115
	Family harmony	152	3	11	166
	Your health	273	6	7	286
Total			543	14	35	592

Source: Data processed from Visauta Vinacua (1997) CD database Statistical Analysis with SPSS for Windows.... File work.sav

Table 33 shows the other part of the tabular segmentation by item division, the reader can corroborate that the totals are maintained.

Table 33. B. Contingency table with category in structure layer $b_2/(_4A\,|_3B)$

Genre b_2: Male					
				B: Religión	
		Catholic	Other	None	Total
A1: Concerns: aspect that worries you the most	Your emotional life	31	2	5	38
	Money	152	5	14	171
	Family harmony	103	1	7	111
	Your health	203	2	27	232
Total			489	10	53	552

Source: Data processed from Visauta Vinacua (1997) CD database Statistical Analysis with SPSS for Windows.... File work.sav

Three-dimensional table with stacked categories.

The following example in table 34 shows a further degree of complexity of the three-dimensional table. The structure of this table is $b_2/(_4A_1\,|_3B)(_4A_2\,|3B)$. Below in the example of table 35 the same information is shown, but using the complement of the division by the item b2 women. The reader should corroborate the totals with the unsplit table 13

Table 34. A. Three-dimensional table with stacked categories (Division by men). $b_2/({}_4A_1\mid{}_3B)({}_4A_2\mid{}_3B)$

Genreb2:Male		B: Religión			
		Catholic	Other	None	Total
A_1 Concerns: aspect that worries you the most	Your emotional life	31	2	5	38
	Money	152	5	14	171
	Family harmony	103	1	7	111
	Your health	203	2	27	232
Total		489	10	53	552
A_2 Concerns: aspect you are least concerned about	Your emotional life	85	2	16	103
	Money	151	3	15	169
	Family harmony	73	3	6	82
	Your health	88	2	9	99
Total		397	10	46	453

Source: Data processed from Visauta Vinacua (1997) CD database Statistical Analysis with SPSS for Windows.... File work.sav

Table 35. Three-dimensional table with stacked categories (Division by women). $b_1/(({}_4A_1\mid{}_3B)({}_4A_2\mid{}_3B))$

Genre b_1: Female		B: Religión			
		Catholic	Other	None	Total
A_1. Concerns: aspect that worries you the most	Your emotional life	22		3	25
	Money	96	5	14	115
	Family harmony	152	3	11	166
	Your health	273	6	7	286
Total		543	14	35	592
A_2. Concerns: aspect you are least concerned about	Your emotional life	100	4	4	108
	Money	218	3	8	229
	Family harmony	75	3	2	80
	Your health	55	4	13	72
Total		448	14	27	489

Fuente: Datos procesados de la base de datos del CD de Visauta Vinacua (1997) Análisis Estadístico con SPSS para Windows... Archivo trabajo.sav

Three-dimensional table with nested categories.

This table 36 has a higher degree of complexity. The three-dimensional tables 34 and 35 would be very large together or stacked, the structure of table 36 is $b_2/[({}_6D/{}_3B)\mid{}_4A_1]$

Table 36. Three-dimensional structure with nested categories (Division by men) *. $b_2/[(_6D/_3B)|_4A_1]$

Genre b2: Male				A1: Concerns: aspect that worries you the most			
				Your emotional life	Money	Family harmony	Your health
				N	N	N	N
D: Estado civil		B Religión	Catholic	13	58	23	50
			Other		1		
			None	4	8	1	15
		Total		17	67	24	65
	Married	B Religión	Catholic	18	89	72	143
			Other	2	4	1	2
			None	1	6	6	12
		Total		21	99	79	157
	Lives with a partner	B Religión	Catholic		2	2	
			Other				
			None				
		Total			2	2	
	Separated	B Religión	Catholic		1	2	3
			Other				
			None				
		Total			1	2	3
	Divorced	B Religión	Catholic		1	1	
			Other				
			None				
		Total			1	1	
	Widowed	B Religión	Catholic		1	3	7
			Other				
			None				
		Total			1	3	7

Source: Data processed from Visauta Vinacua (1997) CD database Statistical Analysis with SPSS for Windows.... File work.sav
(*): Blank cells indicate absence of data in that combination.

These blank cells indicate the effect of the increase of the *Intension* with the consequent decrease of the *Extension*. It remains for the reader to find the structure of 37, how would this table admit a segmentation? Note that it is already segmented so it must be segmented from the original in another way.

Table 37. Three-dimensional table with nested categories (Division by women)

Genre 2 Female				A$_1$ Concerns: aspect that worries you the most			
				Your emotional life	Money	Family harmony	Your health
				N	N	N	N
D. Marital status	Single	B. Religión	Catholic	8	33	26	36
			Other	...	...	1	...
			None	1	8	7	3
		Total		9	41	34	39
	Married	B Religión	Catholic	9	59	112	182
			Other		4	2	5
			None	2	3	2	1
		Total		11	66	116	188
	Lives with a partner	B Religión	Catholic	...	...	...	1
			Other	...	...	...	...
			None		3	...	1
		Total		...	3	...	2
	Separated	B Religión	Catholic	2	1	1	1
			Other	...	1	...	...
			None	...	...	...	1
		Total		2	2	1	2
	Divorced	B Religión	Catholic	...	...	3	...
			Other	...	...	...	...
			None	...	...	...	...
		Total		...	...	3	...
	Widowed	B Religión	Catholic	3	3	10	53
			Other	...	...	...	1
			None	...	...	2	1
		Total		3	3	12	55

Source: Data processed from Visauta Vinacua (1997) CD database Statistical Analysis with SPSS for Windows.... File work.sav
(...): Absence of data in that combination.

Three-dimensional table with nested and stacked categories.

This last table is more complex and larger as it has layer, stacked and nested categories. The following tables 38 and 39 show two tables in which the complete table has been segmented because it is very large. The null triple combinations or frequencies have been eliminated, giving the appearance of a small table, and some totals have also been eliminated due to the existence of a single summand, since at this point we are not interested in the interpretation but in the presentation of the complexity of the table and the notation of the structure, which in this case is $b_1/[(_6D/_3C)|A_1][(_6D/_3C)|A_2]$

Table 38. Three-dimensional structure with nested and stacked categories (Division b1. Male)

Genre b₁.Male		A₁ Aspect of greatest concern					A₂ Aspect of least concern				
D Marital Status	C Religión	Your emotional life	Money	Family harmony	Your health	Total	Your emotional life	Money	Family harmony	Your health	Total
Single	Catholic	13	58	23	50	144	31	32	24	32	119
	Other	*	1	*	*	1	1	*	*	*	1
	None	4	8	1	15	28	12	6	4	3	25
	Total	17	9	24	55	173	44	38	28	35	145
Married	Catholic	18	89	72	143	322	49	108	46	54	257
	Other	2	4	1	2	9	1	3	3	2	9
	None	1	6	6	12	25	4	9	2	6	21
	Total	21	99	79	157	356	54	120	51	62	287
Lives with a partner	Catholic	*	2	2	*	4	1	2	*	1	4
Separated	Catholic	*	1	2	3	6	1	2	2	*	5
Divorced	Catholic	*	1	1		2	1	*	*	*	1
Widowed	Catholic	*	1	3	7	11	2	7	1	1	11

Fuente: Datos procesados de la base de datos del CD de Visauta Vinacua (1997) Análisis Estadístico con SPSS para Windows... Archivo trabajo.sav

(*): Las celdas en blanco indican ausencia de datos en esa combinación

The example in Table 39 shows the other value of Division Segmentation b1. Female. The table is simplified by eliminating the null f3 combinations or frequencies, i.e., note that the Marital Status categories do not show all the nested Religion categories (Lives with a partner and Divorced), since they are null (they are zeros). Also, note that the name of the categories Marital Status and Religion are not behind the items as in all the previous ones, this is due to the size of the table, but it is evident that Marital Status nests Religion. It is up to the reader to find the structure

Table 39. Three-dimensional structure with nested and stacked categories (Division b1. Female)

Genre b$_1$.Female		Aspect of greatest concern					Aspect of least concern				
Marital Status	Religión	Your emotional life	Money	Family harmony	Your health	Total	Your emotional life	Money	Family harmony	Your health	Total
Single	Catholic	8	33	26	36	103	24	32	19	16	91
	Other	*	*	1	*	1	*	*	*	1	1
	None	1	8	7	3	19	2	4	2	6	14
	Total	9	41	34	39	123	26	36	21	23	106
Married	Catholic	9	59	112	182	362	67	143	47	34	291
	Other	*	4	2	5	11	3	3	2	3	11
	None	2	3	2	1	8	1	1	*	5	7
	Total	11	66	116	188	381	71	147	49	42	309
Lives with a partner	Catholic	*	*	*	1	1	*	*	1	*	1
	None	*	3	*	1	4	1	*	*	2	3
	Total	0	3	0	2	5	1	0	1	2	4
Separated	Catholic	2	1	1	1	5	1	2	1	*	4
	Other	*	1	*	*	1	1	*	*	*	1
	None	*	*	*	1	1	*	1	*	*	1
	Total	2	2	1	2	7	2	3	1	0	6
Divorced	Catholic	*	*	3	*	3	1	1	*	*	2
Widowed	Catholic	3	3	10	53	69	7	40	7	5	59
	Other	*	*	*	1	1	*	*	1	*	1
	None	*	*	2	1	3	*	2	*	*	2
	Total	3	3	12	55	73	7	42	8	5	62

Source: Data processed from Visauta Vinacua (1997) CD database Statistical Analysis with SPSS for Windows.... File work.sav
(*): Blank cells indicate absence of data in that combination.

Chapter 5. Manual table construction

Manual construction through exercises

The following database (Table 40) contains the variables sex and educational level of 20 individuals. The sex category has been labeled with the values {1= Female, 2= Male} and Education with {1= Primary, 2= Secondary, 3= University}. The study was conducted by means of surveys in the 23 de enero parish of Caracas in March 2007. With this information, the following is requested.

Table 40. Listing of sex and education in 20 individuals

Individual	Sex	Education	Individual	Sex	Education
1	1	1	11	2	2
2	1	2	13	1	1
3	2	1	14	1	1
4	2	2	15	1	2
5	2	2	16	2	3
6	2	3	17	2	2
7	1	2	18	2	3
8	2	3	19	2	2
9	1	3	20	2	3
10	2	3			

1. Exercise1: Make two simple SS tables for sex and education using the frequency and percentage statistics.

2. Exercise2: Place the title of the table, the labels and values of the categories, those of the statistics and the table order.

3. Exercise 3: Read the results.

4. Exercise 4: Make a stacked table SL for the two variables by repeating point 2 and 3.

5. Exercise 5: Nest Sex Education in a SS table repeating items 2 and 3.

6. Exercise 6: Nest Sex over Education in a SL table repeating items 2 and 3 (to be done by the student).

Solution to Exercise 1: Simple Tables SS

We will make the simple sex table, the student will make the education table. We will start by designing the table. As it is SS we ask for the labels and values of the categories at the top and the statistics at the top as well. For Sex we will have the following design scheme, one cell for the label. Below it two for the values (male and Female) and 1 for the total in position S. Then, also in position S, we will have two cells for "n" and "%" below each sex value. Let's see the design

1. Cell for tag Gender

Gender

2. Insert another row and divide it into three segments, one for male, one for Female and one for total.

Gender		
Male	Female	Total

3. Below each cell we need two cells for "n" and "%". We insert another line and divide it into 6 cells

Gender					
Male		Male		Male	
N	%	N	%	N	%

4. Finally we insert another line with six cells below for the body of the table

Gender					
Male		Male		Male	
N	%	N	%	N	%

5. We have finished the design of the table. We now count the male (1) and Female (2) values and the total

Gender					
Male		Male		Male	
N	%	N	%	N	%
7	35	13	65	20	100

Finally, we obtain the % of each value with respect to the total: 7/20 *100 = 35% and 13/20*100 = 65%.

Exercise 2: Table title and table footer

Table nº ( ) Sex of the inhabitants of the parish 23 de Enero de Caracas. Frequencies and percentages (Total base of interviews).

Gender					
Male		Male		Male	
N	%	N	%	N	%
7	35	13	65	20	100

Source: Interviews conducted with 20 inhabitants of the parish of 23 de enero in the city of Caracas during the month of March 2007.

Note that the title describes the content of the table and that the base on which the percentages have been obtained has been added, i.e. if 20 interviews were carried out, the base total interviews is added.

Exercise 3: Reading values:

Values are read under the grammatical structure of subject and predicate. The subject in this case is the number and we will call it the statistical subject and the predicate is that which is said to have that number. The *statistical subject* must be accompanied by a complement that defines it. The classic form is:

the (percent) of the (comparison base or total) is (category label). This would leave the (statistical subject) of the (complement) is (predicate). Thus, for example, 35% of the total number of respondents is male; 65% of the total number of respondents is Female. It is a mistake to say that 35% of the totals ARE male, since the verb BE refers to the singular of percentage.

Exercise 4: Stacked table SL

As stacking is placing two simple tables together we will do the calculation of the cells separately. Recall that the table is SL, putting the statistics sideways which means adding a cell to the left for them.

1. We insert a row, divide it into three cells (one for the L position of the statisticians and two others for sex and education) and add the category labels, leaving the first cell blank.

	Sex	Education

2. Since sex has two values[8] and education has three, we insert another row and divide it into 8 cells (position L, male, Female, total sex; primary, secondary, university, total education).

	Sex			Education			
	Male	Female	Total	Elementary School	High School	University	Total

3. Now we add two rows the labels of the statistics (n and %) in position L

	Sex			Education			
	Male	Female	Total	Elementary School	High School	University	Total
N							
%							

4. We add the statistical analysis or counting of the category label values and obtaining the percentages.

	Sex			Education			
	Male	Female	Total	Elementary School	High School	University	Total
N	7	13	20	5	8	7	20
%	35	65	100	25	40	35	100

Title and Footer

Table n° ( ). Sex and educational level of the inhabitants of the parish of 23 de enero de Caracas. Frequencies and percentages (Total base of interviews).

	Sex	Education

---

[8] We distinguish between gender as a sexual preference and sex as a natural genital apparatus.

	Male	Female	Total	Elementary School	High School	University	Total
N	7	13	20	5	8	7	20
%	35	65	100	25	40	35	100

Source: Interviews conducted with 20 inhabitants of the parish of 23 de enero in the city of Caracas during the month of March 2007.

Reading of values

We will read only one value. 40% of the total number of respondents in the parish of 23 de enero in Caracas have a high school education.

Exercise 5: Nesting sex education in a SS table

1. Since they ask for a nested table, education over sex, we must design the table so that the education labels nest the sex labels. That order will be education → education values → sex → sex values, in S position. We insert a row for education

Education

2. Next we add another row and divide it into three cells for each category

Education		
Elementary School	High School	University

3. Now let's nest, add another row and divide it into 6 cells, for the sex label and the total

Education					
Elementary School		Elementary School		Elementary School	
Sex	Total	Sex	Total	Sex	Total

4. Now we add the gender values plus the total, that is, three cells per group for primary, secondary and university.

Education								
Elementary School			Elementary School			Elementary School		
Sex		Total	Sex		Total	Sex		Total
Male	Female		Male	Female		Male	Female	

5. Now we add another row for the statistics in position S, i.e. 2 cells (n and %) for each sex value and two for each total (n and %).

Education																	
Elementary School						Elementary School						Elementary School					
Sex				Sex		Sex				Sex		Sex				Sex	
Male		Male		Male		Male		Male		Male		Male		Male		Male	
N	%	N	%	N	%	N	%	N	%	N	%	N	%	N	%	N	%

6. Finally we add the row for the table body

Education																	
Elementary School						Elementary School						Elementary School					
Sex				Sex		Sex				Sex		Sex				Sex	
Male		Male		Male		Male		Male		Male		Male		Male		Male	
N	%	N	%	N	%	N	%	N	%	N	%	N	%	N	%	N	%

Now we count the data. Recall that we must count double frequencies in the following order: Primary-Male; Primary-Female or values 1-1 and 1-2. Then Secondary-Male and Secondary-Female or values 2-1 and 2-2 and so on. Let's see 1-1 (Primary-Male) there are 3 (individuals 1, 13 and 14); 1-2 (Primary-Female) there are 2 (individuals 3 and 12) and so on.

Education																	
Elementary School						Elementary School						Elementary School					
Sex				Sex		Sex				Sex		Sex				Sex	
Male		Male		Male		Male		Male		Male		Male		Male		Male	
N	%	N	%	N	%	N	%	N	%	N	%	N	%	N	%	N	%
3	60	2	40	5	100	3	37,5	5	62,5	8	100	1	14,29	6	85,71	7	100

The percentage should be obtained (in this case) over the total group (primary, secondary and university).

Title and Table Footer

Title: Table n° ( ) Education by sex of the inhabitants of the parish 23 de Enero de Caracas. Frequencies and percentages (Base: total per group).

Foot of Table: Source: Source: Interviews conducted with 20 inhabitants of the 23 de Enero parish in the city of Caracas during the month of March 2007.

Reading of the cells:

Since the table is nested, the percentage is taken over the total group, therefore the statistical subject (percentage) has as a complement the group that is nested and the predicate is the group that is nested since it is the one from which the percentage is taken.

The first would be as follows: 60% of the inhabitants who have primary education are male. The fifth percentage of the table would read as follows: 62.5% of the inhabitants with secondary education are Female. Complete the other percentages

Exercise 7: In the previous nesting exercise the percentages were taken on the group totals. But this option is not the only one, you can also take the percentages on the table total and have the percentages on the group total and the table total. Essentially the table remains the same only that the cells of the table total would have to be added to it. Let's see how to do it for a SL nesting (to save space)

Exercise 7

1. We start by inserting the row for education and divide it into three cells, the statisticians' cell in position L, the label cell and the total table cell.

	Education	Total

2. Then we insert another row and divide it into 5 cells, the one for the statistics in position L, three for the education values and the one for the total.

	Education			Total
	Elementary School	High School	University	

3. Now we nest by inserting a row and divide it into two cells per education group, one for gender and one for total group. There are 8 cells, two per group, the statistics and the total.

Education						Total Tabla
Elementary School		High School		University		
Sex	Total Group	Sex	Total Group	Sex	Total Group	

4. We insert another row and divide each gender label into two cells (male and Female). This gives 11 cells

	Education									Total Table
	Elementary School			Elementary School			Elementary School			
	Sex		Total Group	Sex		Total Group	Sex		Total Group	
	Male	Female		Male	Female		Male	Female		

5. We add the row where the statistics labels go. There are three rows with the same number of cells, one for the frequencies and two for the percentages per group and one for the total.

	Education									Total Table
	Elementary School			Elementary School			Elementary School			
	Sex		Total Group	Sex		Total Group	Sex		Total Group	
	Male	Female		Male	Female		Male	Female		
N	3	2	5	3	5	8	1	6	7	20
% Group	60	40	100	37,5	62,5	100	14,29	85,71	100	
% Total	15	10	25	15	25	40	5	30	35	100

6. The count and percentages for nesting are similar to the previous exercise. The count for the total changes since it is total per group. Notice in the table of exercise 1 how the cells for the frequencies are similar, the cells for the percentages change with respect to the total.

Exercise 8: The following table 41 shows 30 values for age and sex of teachers at a regional university in Venezuela (UNET) coded as follows. Age {1: 18 to 25; 2: 26 to 35: 3: 36 to 55; 4: 56 to 65}; Sex {1= Female; 2= Male}. Note that there is a missing data (missing value). You are asked to

Table 41. Age and sex listing for 30 individuals

Case	Age	Sex	Case	Age	Sex
1	1	1	16	3	2
2	1	1	17	2	2
3	1	1	18	2	1
4	2	1	19	1	2
5		2	20	2	2
6	2	2	21	2	1
7	1	2	22	3	2
8	1	2	23	4	2
9	1	2	24	4	2
10	1	2	25	2	2
11	1	2	26	2	1
12	1	1	27	3	2
13	1	1	28	3	2
14	2	1	29	2	2
15	4	2	30	3	2

6.1. Construct a double-entry table (age in the rows and sex in the columns) (also called a double-entry table with sex SL and age LL) Why?

6.2. Count the double frequencies of the variables and place them in the table.

6.3. Obtain the row percentage

6.4. Place a table title and a table footnote (stating the missing value).

Exercise 8

As Sex goes in SL position and Age in LL position, it indicates that the labels of the statistics go in lateral position (this is always the case in double-entry tables). Now, in position S go the labels of the categories, the values of Sex (male and Female) and the total that would give 4 columns; however, space must be saved for Age that goes in position L. That is, a column for the values of age and another for the labels of the statistics, would be 2 and in total of 6 columns.

On the other hand, the Age category has 4 values (4 rows), additionally we have to count the row of totals, which accumulate 5 rows (additionally a row for N and another for % row for each value and total), it will be 5 * 2 = 10 rows. To this is added the one for the name of the sex category and another for its values (2) giving 12 rows. A matrix of 12 rows by 6 columns

1. Matrix calculation (we calculate it differently from the double-entry table section) 2.

2. Table design. We place the name of the column variable (Sex) at the beginning of the second column and Age in the third cell of the first column.

	Sexo				
Edad					

3. Table layout. We place from right to left (in the second line at the height of Age) the total and the values of sex (in reverse order to how we see them)

	Sexo		Femaleenino	Maleulino	Total
Edad					

4. Next in front of Age we place the values of the category every three cells (leaving space for N and %) and total

	Sex		Female	Maleo	Total
Age	From 18 to 25				
	From 26 to 35				
	From 36 to 55				
	From 56 to 65				
	Total				

5. We add N and % in each age group and total.

	Sex		Female	Maleo	Total
Age	From 18 to 25				
		% frow			
	From 26 to 35	N			
		% frow			
	From 36 to 55	N			
		% frow			
	From 56 to 65	N			
		% row			
	Total	N			
		% row			

6. Counting values. We start counting the cell Female-From 18 to 25 years corresponding to the codes 1=Female and 1=From 18 to 25. We search the database for the pairs (1,1) = (Female, From 18 to 25) being 5 and repeat the process for cells (1,2); (2,1); (2,2); (3,1); (3,2); (4,1); (4,2) [we have placed the values to be counted in the cells for the student to do so].

	Sex				
			Female	Maleo	Total
Age	From 18 to 25		5 (1,1)	(1,2	
		% frow			
	From 26 to 35	N	(2,1)	(2,2)	
		% frow			
	From 36 to 55	N	(3,1)	(3,2)	
		% frow			
	From 56 to 65	N	(4,1)	(4,2)	
		% row			
	Total	N			
		% row			

7. 7. Then the row and column totals are obtained; the percentages per row are calculated and the desired style is elaborated giving the following table

			Sex		Total
			Female	Male	
Edad	Frome 18 to 25	N	5	6	11
		% row	45.5	54.5	100
	From 26 to 35	N	5	5	10
		% row	50	50	100
	From 36 to 55	N		5	5
		% row		100	100
	From 56 to 65	N		3	3
		% row		100	100
	Total	N	10	19	29
		% row	34.5	65.5	100

Unsolved exercises

1. Exercise Suppose you want to make a table to present the number of students per section {1A, 1B, 1C} of La Buena Marcha High School. You also want to present the grades that were classified as {Good, Fair and Deficient]. The simple frequency and percentage statistics are used. With this information make the following tables

1.1. A simple table SS for sections

1.2. A simple table SL for the sections

1.3. One table LL for the sections

1.4. A table LS for the sections

1.5. Repeat the previous points by making the same tables but for the notes.

2. Exercise Suppose you want to make a table to present grades and grades stacked and combined. Make the following tables

2.1. A simple SS table for the sections by note

2.2. A simple table SL for the sections by note

2.3. A table LL for the sections per note

2.4. A table LS for the sections by note

2.5. Repeat the previous points by making the same tables but for notes per section.

2.6. In these cases would you use single or double frequencies? Why?

3. Exercise: From the same case as above, we now ask you to incorporate the sex of the students both by section and by grades. Make the following tables

3.1. Stacked SS for notes, section and gender

3.2. Nested SL section on sex

3.3. Nested SL gender on section

3.4. Nested SS notes on gender

3.5. Nested SS section over sex notes (which is equivalent to using section as layer)

4. Exercise: The dental clinics decide to promote the oral health of their patients by giving talks on oral health and teaching proper brushing. Clinics in two parishes (parishes A and B) participate in the program. An investigation on the activity of the clinics in 15 days shows the following results: A, 35, B, 28, A, 41 and B, 32. With this information, construct the following tables

4.1. Stacked table SS total of talks and brushing instruction of the clinics

4.2. Stacked table SL total of clinic brushing lectures and instruction

5. 5. Exercise: The following database shows the variables sex, age, year of graduate studies and specialty for 39 students of the graduate program in dentistry at the Central University of Venezuela (the data have been altered to summarize them). With this information the following is requested

Table 42 Data on 39 postgraduate dental students. Sex, age, year completed and specialty

N°	Sex	Age	year attended	Specialty	N°	Sex	Age	year attended	Specialty
1	Female	26	Second	Orthodontics	21	Female	27	First	Endodontics
2	Female	28	Second	Oral Surgery	22	Male	30	First	Endodontics
3	Female	27	First	Oral Surgery	23	Female	26	First	Orthodontics
4	Female	26	Second	Endodontics	24	Male	26	First	Oral Surgery
5	Female	36	First	Endodontics	25	Female	26	First	Oral Surgery
6	Female	24	First	Oral Surgery	26	Female	38	Second	Orthodontics
7	Female	25	Second	Orthodontics	27	Female	26	Second	Oral Surgery
8	Male	28	Second	Endodontics	28	Female	26	First	Endodontics
9	Female	27	Second	Oral Surgery	29	Female	29	First	Orthodontics
10	Male	30	First	Endodontics	30	Female	29	Second	Orthodontics
11	Female	27	First	Orthodontics	31	Female	28	Second	Oral Surgery
12	Male	28	Second	Orthodontics	32	Female	38	First	Orthodontics
13	Female	31	First	Oral Surgery	33	Male	30	Second	Orthodontics
14	Female	27	First	Oral Surgery	34	Female	26	Second	Endodontics
15	Female	28	Second	Oral Surgery	35	Female	29	Second	Oral Surgery
16	Male	36	First	Oral Surgery	36	Female	25	Second	Orthodontics
17	Female	27	Second	Orthodontics	37	Female	26	First	Endodontics
18	Male	28	First	Orthodontics	38	Male	26	First	Orthodontics
19	Female	39	Second	Orthodontics	39	Male	24	Second	Endodontics
20	Female	29	Second	Endodontics					

5.1. Code the categories sex, year of study and specialty.

5.2. Group the age in 3 classes. The first one from 24 to 27, the second one from 27 to 29 and the third one from 30 or more.

5.3. Construct a stacked table SL sex, year of study and specialty obtaining the statistics absolute frequency and percentage.

5.4. Construct a nested table sex over specialty with N and % group.

5.5. Construct a nested table of specialty over sex with N and % group.

5.6. With the above information construct a double-entry table with sex in the columns and specialty in the rows. Obtain the frequencies and percentages

5.7. Construct a nested table age group over specialty. Use N and % group and % total

6. Exercise: The following table 43 shows 33 interviews with visitors to a shopping mall in the city of Caracas. The variables are Reason for visiting the shopping center {1: Works there; 2: Goes for another reason}; Sex {1: Male;

2: Female}; Municipality of origin {1: Libertador; 2: Chacao, 3: Baruta, 4: Another place}; Place of origin {1: From home; 2: From another place}. With this information obtain

6.1. Simple absolute and relative frequencies of each category or variable.

6.2. Stack the variables Reason for visit and Place of origin; Sex and Place of origin; Reason for visit and Sex in two separate tables SS and SL.

6.3. Double frequencies of Reason for visit and Place of Origin; of Sex and Place of Origin; of Reason for visit and Sex.

6.4. With the previous information, obtain the nestings of Reason for visit over Place of origin; Sex over Place of origin; Reason for visit over Sex and obtain the percentages.

6.5. Now make the contingency or double entry tables of Reason for visit VS Place of Origin; Sex VS Place of Origin; Reason for visit VS Sex and obtain the row percentages and separately the column percentages (Say to which nesting do these percentages correspond?).

Tabla 43 List of the 33 interviews and the 4 categories

Reason for visit	Sex	Municipality of origin	Place of origin
2	1	1	1
2	1	1	2
2	1	1	1
1	1	1	1
1	1	1	2
1	1	1	1
2	1	4	2
2	1	4	1
2	1	4	2
2	1	4	1
2	1	4	1
2	1	4	1
1	2	1	1
1	2	1	1
2	2	2	1
2	2	1	1
2	2	1	2
2	2	1	2
2	2	1	1
2	1	1	1
2	1	4	1
2	1	4	2
2	1	4	2
2	1	1	1
2	1	1	2
2	2	3	1
2	2	3	1
1	2	1	1
1	2	4	1
1	2	1	1
2	1	1	2
2	2	1	2
2	2	4	2

PART THREE

Tabular Dynamics

Introdución

Methodologically, the statistical table was always seen as the final result of data processing, they were self-understandable tables as they consisted of a visual arrangement of the results, although some were extensive due to the number of cases offered as lists by region or product. With the development of software, these processing summaries can be presented in different ways and complexity to make efficient the two-dimensional space of the sheet, besides allowing to build rows and columns easily, therefore, the statistical table acquires characteristics of analysis method allowing the design of the *Basic Tabulations Plan*, which is an element of the statistical design for a sample. This plan was difficult because it was manual and correcting it implied redoing the tables manually, this was left behind, but the tabulation plan was practically forgotten, nowadays researches are done paralyzed in data processing due to the non-existence of a tabulation plan that indicates how the variables will be analyzed, how the crossings will be, etc., therefore, it is necessary to design a tabular logic that addresses these issues.

A logic as a set of symbols, construction rules and a semantics of interpretation has several objectives, in our case we have pointed out a first objective of communication of the tabular structure because there is no linguistic system to convey a requirement or its characteristics therefore the teacher-student, researcher-researcher interaction is limited preventing to close the existing technological gap, A second objective has to do with the usability of the tables, those that are large have problems of presentation on the screen being illegible many times and a third objective is that beyond the ergonomic problems the statistical table is a system of composition of variables that shows a research design, that is a table must allow the construction of a discourse of interpretation. All this is facilitated by the tabular dynamics in the computer and the construction programs.

This third part has three chapters in the First (sixth of the numbering) we deal with the definitions of the tabular logic or way of thinking the tables, we arrive at a classification of the tables not according to their dimensionality but according to their configuration, in chapter 7 we study the method of *semantic reduction* that allows the obtaining of the accumulation of totals on which it is possible to relativize, in this chapter is included the elaboration of the Basic Tabulations Plan as concretion of a quantitative investigation and in chapter 8 the way of analyzing a table depending on the type of table is shown.

Chapter 6. Semantic or Qualitative Tabular Analysis

Predicate Intension and Extension

Statistical sampling has two components, the quantitative one consists of defining a Universe or total of elements from which a part will be taken to be examined, and the qualitative one consists of the attributes to be considered in the respective Unit of Analysis, whether it is the dwelling, the person or a part of the person, companies, among other units that have their own characteristics defined in the research plan. These quantitative and qualitative elements respond to the first order logic that is the language used to construct the definitions of the samples. Thus the unit of analysis contains the total of characteristics to be evaluated and that represent in logic the *intension* or connotation, while the universe or total of elements that possess such characteristics represent the *extension* or denotation[9].

In sampling terms, intention implies *vertical coverage* or levels of disaggregation of the variable, for example, the educational level variable is disaggregated into primary, basic and university, and can also be primary, middle, university and post-university; extension implies horizontal or total coverage of elements to be studied in a given area or areas. For example, the term "inhabitant of Argentina" as a class increases its intension with the addition of attributes such as "inhabitant of Argentina who lives in the capital" and increases with "inhabitant of Argentina who lives in the capital and is a man". When increasing the intension generally (not always) decreases the extension because it is assumed that there are more Argentines living in the country than those living in the capital.

In statistical tables we have a particular case of intensionality since the samples can include several classes since the unit of analysis may contain several, for example, an elementary school, where there are teachers, students, janitors, administrative staff, as well as the types of furniture and all this can be included in the tables to be analyzed, therefore, we will define the sample intension and extension.

Sampling intent

The *sample intent* is the totality of the variables with their items or values that are going to be analyzed in the tables, the intent is shown in the complete headings in such a way that the percentages of frequencies are based on a totalization and this is always a reduction of predicates by way of reduction of the intent by elimination of characteristics and we call it *semantic reduction*. That is, every classification or subset belongs to a larger class by reduction of intent as we saw. In our case holons represent sets within larger sets that are totalized by eliminating (counting) the elements of the sets.

---

[9] For these concepts see, among others, Irving Copy and Carl Kohen (2013).

Cuadro 8. Structure B|A

B|A		A		
		a_1	a_2	Total
		N	N	N
B	b_1	b_1a_1	b_1a_2	b_1
	b_2	b_2a_1	b_2a_2	b_2
	Total	a_1	a_2	TT

For example, Table 10, B|A, shows the semantic b_ja_i; the semantic reduction of $b_1a_1+b_1a_2 = b_1(a_1+a_2)$ or simply b_1 *independientemente de* a_i, therefore, we write $= b_1$ without loss of logical significance (moreover, we omit the predicate variables xb_1a_1, for simplicity, as we focus on the semantics); that is, by counting the conjugate items, the semantics a1 and a2 are hidden in the total; the intensionality of each frequency has been reduced by the total. Example, if $b_1 =$ men, $b_2 =$ women; $a_1 =$ high, $a_2 =$ low, and we have 5 b1a1 and 10 b_1a_2, the total is 15 b_1 (a_1 or a_2). It can be seen that the Second holon has b_2 as its total, which when totaled with b_1 results in the extension of all the sets (TT); likewise with a1 and a_2. These operations are justified in predicate logic by means of the union and intercept of predicates

Thus a semantic or combination of predicates accompanying variable x is a mirror of the tabular configuration and is composed of two terms, the combination of predicates in position S and in position L; for example (a1b1)(c1d1) is the mirror of the configuration A|B if both variables had two items each. In this case the Firsts elements of each term is a layer item because A is a nesting variable of B and C is a nesting variable of D.

Sample extension

The process of semantic reduction has as a consequence the counting of the values of the items in the totals and this accumulation goes from the conjugated items to the layers and so on, a process that we will see happens by rows and columns until reaching the total total of the sample. Now as the reduction means to eliminate the semantics of the items or values in the count or grouping, when reaching the total total, all the values or semantics will have been reduced leaving only one number, the wrong saying that pears and apples cannot be added together does not make sense because it reflects a semantic reduction, 4 pears + 5 apples = 9 fruits, that is, pears and apples go to the class fruits, that is, an indistinguishable generic because when saying 9 fruits the individuality of the data is lost because of the class, that is, if we say 9 fruits we will no longer know if they are pears, apples or grapes, so that this total total is *mute*, generic, it cannot refer to any element of the universe of discourse, it is absolute. This makes sense because when we say that a city has a census of 2.5 million people, we know nothing about them because of the number, they are all xi, any one that has any characteristic, while an x1 is already a particular one. So the grand total or total total total is mute in the sense that it only refers to the cardinal of the universe.

Now let us see how the semantic reduction proceeds.

Reduction of intension by layers and calculation of own totals

Box 9 shows the tabular structure $\Phi = (_2A/_2B) \mid (_2C/_2D)$. Variable A nesting to B, in position 1 and C nesting to D, in position s. This configuration presents semantics $a_ib_jc_kd_l$, configured in two conjugate terms $(a_ib_j)(c_kd_l)$. We will distinguish the *proper totals* (TP) from the *subtotals* (ST) or the grand total (TG), the TP are cells of totals included in the ST, that is the structure Φ has 2 subtotals of A and 3 of B due to the nesting and the totals of nested B, for its part C has likewise 2 totals and 3 of D, so that in total there are 10 ST, actually the totals of the nested variables are totals of totals, but we call them all ST to avoid naming complications (as totals of 1st order, 2nd and 3rd order), the TPs are the ones from which the frequencies are relativized.

We will take the first semantic (a_1b_1) (c_1d_1) as a reference for the analysis.

Box 9. Structure $\Phi= (_2A/_2B) \mid (_2C/_2D)$

A	a	B	b	c₁ D d₁	c₁ D d₂	c₁ T	c₂ D d₁	c₂ D d₂	c₂ T	Total D d₁	Total D d₂	Total Total
				N	N	N	N	N	N	N	N	N
	a_1	B	b_1	$(a_1b_1)(c_1d_1)$	$(a_1b_1)(c_1d_2)$	$(a_1b_1)c_1$	$(a_1b_1)(c_2d_1)$		→	$(a_1b_1)d_1$	→	a_1b_1
			b_2	↓		↓				↓		↓
			Total	$a_1(c_1d_1)$	→	a_1c_1			→	a_1d_1	→	a_1
	a_2	B	b_1									
			b_2									
			Total	↓		↓				↓		↓
	Total	B	b_1	$b_1(c_1d_1)$	→	b_1c_1			→	b_1d_1	→	b_1
			b_2	↓		↓				↓		↓
			Total	c_1d_1	→	c_1			→	d_1	→	TT

Note that each table (holon in the structure) has its closing total of the items that compose it and this total closes the nest; the semantic (a_1b_1) (c_1d_1) sequentially eliminates its items in the totalization; these are its *own totals* or totals of the nests since none has an item that is not its own, the only exception being the Grand Total or TT in which all the semantics will have been reduced remaining as absolute value since the count has absorbed them. As the tabular structure is made up of intersecting tables, these form nests whose limits are their own totals; the number of own totals of each item is

$$TP = \sum_{i=1}^{k} \binom{n}{k} = \frac{P_{k,n}}{k!} = \frac{n!}{k!(n-k)!}$$

donde n es el número de ítems de ambos términos, k son las formas en que se van a combinar, secuencialmente desde 1 hasta k items; de manera que *TP= 15*, en este caso. Veamos First el cálculo y luego el proceso de reducción

where n is the number of items of both terms, k are the ways in which they are to be combined, sequentially from 1 to k items; so that *TP= 15*, in this case[10] . Let's see First the calculation and then the reduction process

$$TP = \binom{4}{1} + \binom{4}{2} + \binom{4}{3} + \binom{4}{4} = 15$$

These are the combinations, *abcd; abc; abd; acd; bcd; ab; ac; ad; bc; bd; cd; a; b; c; d.* Note that the order is important because even though the combination abc in quantitative terms is commutative with *bac* o *cba* it indicates nesting in layers[11]. Let us now look at the process of reduction by layers. The reduction proceeds thus, from the term to reduce (a_1b_1) (c_1d_1) horizontally the last item of the second term is eliminated and vertically the last item of the first term is eliminated.

1.   The *semantic value* (a_1b_1) (c_1d_1) is reduced thus, $(a_1b_1)(c_1d_1)+(a_1b_1)(c_1d_2)= (a_1b_1)c_1(d_1$ o $d_2)$ and omitted in the total, i.e. d_1 y d_2 are hidden in the total leaving the semantic (a1b1)c1. Note that we have reduced the layer c1, now in the next total it is reduced like this

2.   Now the value $(a_1b_1)(c_1d_1)$ of layer c_1 is summed with $(a_1b_1)(c_1d_2)$ of layer c_2, leaving semantic $(a_1b_1)d_1$ and repeated with the following one

3.   Finally $(a_1b_1)d_1+ (a_1b_1)d_2$ is totaled, leaving (a1b1) where layers D and C will have been reduced. This process is repeated for the following $(a_1b_1)(c_1d_2)$, in terms of numerical values it is similar to adding 4 values of the boxes or 2 subtotals, but we are interested in the semantic ones.

4.   This process is verified vertically until reaching the total reduction of the semantics in the grand total or mute absolute value.

5.   The reduced values will be recovered in the calculation of the percentages.

Characteristics of semantic reduction

1.   Obtaining numerical totals results in the reduction of intension by extension as we have seen.

---

[10] Here is where we observe that some programs only obtain 5 or 7 useful percentages for tables with a maximum of 3 variables, but that are preserved when increasing the size of the structures. They are based on a fixed structure of totals and with surprising names.

[11] This reduction process becomes more complex for larger structures due to the incorporation of intermediate layers such as layer B in the structure $(_2A/_3B/_3C)|(_3D/_4E)$, due to the need to obtain intermediate sums of subtotals that make the tables inoperative as we will see, here we only show a basic structure for didactic reasons, but also to obtain the proper totals, the structure of these totals is known

2. These totals by reducing semantics become proper totals (TP) which are nothing but a passage of simple elements to a larger class.

3. The reduced semantics are not hidden but omitted in the larger class, for example, five grapes plus three pears are eight fruits and the identity of these elements will be recovered in the relativization of the percentage which is the proportion of grapes and pears in the class Fruits.

4. The proper totals (TP) trim the nest and allow the structuring of the table in different configurations because they contain only semantics of the nest.

5. As the internal nests do not exhaust all the semantics, they have not all been reduced, the reduction continues by creating larger external nests making the table holistic, in other words, the reduction operates in layers creating levels of analysis.

6. The reduction ends by exhausting all semantics in the grand total which is an absolute value referring to all semantics involved.

Nest Formation and Totals

In Box 10 we have summarized the layers of the Σ structure in Table 13 having the first semantic $(a_1 b_1)(c_1 d_1)$ as a reference. From the Σ structure we can observe that $(_2A/_2B) \mid (_2C/_2D)$ has 15 TPs according to the formula of proper totals that can be seen in Box 11, this shows that there are first order totals (a_1, b_1, c_1, d_1), Second as $a_1 d_1$ and third as $a_1(c_1 d_1)$, the grand total or TT is of zero order as it has become an absolute value.

Box 10. Reduced or total semantics proper for the first value of the structure Σ

$(a_1 b_1)\ (c_1 d_1)$	$(a_1 b_1)\ c_1$	$(a_1 b_1)\ d_1$	$a_1 b_1$
$a_1(c_1 d_1)$	$a_1 c_1$	$a_1 d_1$	a_1
$b_1(c_1 d_1)$	$b_1 c_1$	$b_1 d_1$	b_1
$c_1 d_1$	c_1	d_1	0

Now from the tabular formula we can see that the tables A and B in position L, $(_2A/_2B)$ have two items each one forming two holons or rows and each one with its total (b_1, b_2, t_b), on the other hand the table A that is a layer of B has two items that are layers of those of B, $a_1/(b_1, b_2, tb)$ and $a_2/(b_1, b_2, tb)$ in addition to their total $(a_1, a_2, ta,)$ these totals of A are reductions of the items of B, now the totals of the layers of A are reduced totals of B and their total is a total-total. Similarly we can say of the tables that are horizontally positioned, C and D, so that the structure Σ has the form of a cross tabulation of variables in layer, thus the forms $_2A$ and $_2C$ form 4 nests bordered by these layer totals as can be seen in Box 13

Box 11 Nests of the structure $\Phi = (_2A/_2B) \mid (_2C/_2D)$

				C								
				c_1			c_2			Total		
				D			D			D		
				d_1	d_2	T	d_1	d_2	T	d_1	d_2	Total
				N	N	N	N	N	N	N	N	N
A	a_1	B	b_1	NEST I			NEST II					
			b_2									
			Total									
	a_2	B	b_1	NEST III			NEST IV					
			b_2									
			Total									
	Total	B	b_1									
			b_2									
			Total									

It can also be seen that the shaded cells represent the structure of totals, 15 for each item, so that two types of analysis are derived, one by holon, from item to some total of its own, and the other by nests. The analysis strategy of the current tables is restricted to the totals of the major or ST layers. Another analysis strategy is between totals because of the large amount that is produced, sometimes more than the information contained in the cells, these relationships are given as ratios between them.

The analysis by nests is justified by the semantic closeness or meanings of the items, since the layers act as dependent variable but it must be done separately because placing the percentages on the totals themselves in the same tabular structure is very cumbersome, therefore we resort to the semantic table that we will see later.

A disadvantage of this method is that the statistical programs do not calculate the percentages by nests or holons but by columns and rows creating tables with percentages by rows or columns, so that the method of analysis by nests will consist of obtaining from these programs the frequencies in a given structure and then in a spreadsheet to obtain the percentages by nest or holon. However, to the disadvantage of these statistical programs we will say that for small tables the percentages per row and column are useful, but not for larger tables where several variables are nested. Another drawback is that, although this can be done using SQL processes, a large majority of researchers do not have these skills.

Let us now look at the interpretation of Semantic Reduction for different table types in increasing order of complexity.

Analysis of the tabular structure of some table types

Structure β= ($_2$A/$_2$B)$_L$

In this structure "L" indicates that the nesting is in vertical position, the nested variable B has two items indicating that it has a total, the nesting variable or upper layer A also has two items, therefore there is a total for each pair of items of B, so $_2$A shows that there are two totals that separate two nests. . on the other hand as the first item is a1b1 the proper totals will be TP = $_2$C$_1$+$_2$C$_2$= 2+1 = 3, remember that these are ordered combinations, a_i, b_j and a_ib_j shown in table 44, the shaded cells and the TP are in a softer shade of gray.

Table 44 Structure β= ($_2$A/$_2$B)$_L$

A	a$_1$	B	b$_1$	N	53
			b$_2$	N	60
			Total	N	113
	a$_2$	B	b$_1$	N	52
			b$_2$	N	50
			Total	N	102
	Total	B	b$_1$	N	105
			b$_2$	N	110
			Total	N	215

Structure M = ($_3$G/$_2$A)$_L$

Note that this structure M in Table 45 has the nesting in position L for better use of space. The arrangement 2A of the nested variable indicates that it has two items and therefore a total for each pair of them and the form 3G shows that the nesting variable has three items, and as this nests to A will have 3 totals, in total will be three nests, those of layers g1, g2 and g3; thus the structure M having a frequency item f_2 giaj, will have a total of 3 TP = $_2$C$_1$+$_2$C$_2$ for each f_3 . the reader will complete the analysis of structure

Tabla 45 Structure M = ($_3$G/$_2$A)$_L$

G				N
	g1	A	a1	39
			a2	31
			Total	70
	g2	A	a1	42
			a2	38
			Total	80
	g3	A	a1	32
			a2	33
			Total	65
	Total	A	a1	113
			a2	102
			Total	215

Structure H = $_2A|(_3G/_2B)$

The tabular structure H in table 46 is composed of three variables or tables in this logic, it is not necessary to specify the forms that are in position L or S since the separator "|" already indicates it; the form in position L has two items without nesting indicating that it has only one total, while the nesting in position S has two items in B nested by 3 of G therefore there are three totals in the nesting S and one in L giving three nests. Since the generic item is $a_i g_j b_k$ the TP = $_3C_1 + _3C_2 + _3C_3 = 7$. Again the reader will complete the analysis of the totals.

Tabla 46 Structure H = $_2A|(_3G/_2B)$

			G												
			g1			g2			g3			Total		
			B			B			B			B		
			b1	b2	Total	b1	b2	Total	b1	b2	Total	b1	b2	Total
	a1	N	18	21	39	23	19	42	12	20	32	53	60	113
A	a2	N	18	13	31	17	21	38	17	16	33	52	50	102
	Total	N	36	34	70	40	40	80	29	36	65	105	110	215

Structure Ψ= $(_2C/_2D/_2E)|(_2A/_2B)$

This structure ψ shown in table 47 is larger and more complex because it is of frequencies f_5 so that the TP structure is of order 4 as can be seen from the generic $c_i d_j e_k a_m b_n$, from the structure L it can be noted that there are 8 rows (see the shape of the structure) and from the shape in position S there are 4 columns, so that there are 32 cells to be analyzed, they are conformed in 8 nests that can also be read in the formula 2 of C by 2 of D by 2 of A. The TP = $_5C_1 + _5C_2 + _5C_3 + _5C_4 + _5C_5 = 5+10+10+5+1= 31$ which can be seen in the following chart (Box)12

Box 12 Summary of the structure's own totals Ψ= $(_2C/_2D/_2E)|(_2A/_2B)$[1]

5C1	5C2	5C3	5C4	5C5
1er level	2do level	3er level	4to level	5to level
a	cd	cab	cdea	cdeab
b	ce	cda	cdab	
c	ca	cdb	cdeb	
d	cb	cde	ceab	
e	de	cea	deab	
	da	ceb		
	db	dab		
	ea	dea		
	eb	deb		
	ab	eab		

1. We omit the subscripts of the semantic ones.

The reader can check the totals for each semantic. We have colored in gray the TP of the first semantic and in gray the subtotals and grand totals.

Table 47. Structure Ψ= $(_2C/_2D/_2E)|(_2A/_2B)$

						A								
						a1			a2			Total		
						B			B			B		
						b1	b2	Total	b1	b2	Total	b1	b2	Total
C		D		E		N	N	N	N	N	N	N	N	N
C	c1	D	d1	E	e1	6	7	13	16	3	19	22	10	32
					e2	6	15	21	7	10	17	13	25	38
					Total	12	22	34	23	13	36	35	35	70
			d2	E	e1	7	6	13	2	8	10	9	14	23
					e2	9	8	17	10	9	19	19	17	36
					Total	16	14	30	12	17	29	28	31	59
			Total	E	e1	13	13	26	18	11	29	31	24	55
					e2	15	23	38	17	19	36	32	42	74
					Total	28	36	64	35	30	65	63	66	129
	c2	D	d1	E	e1	5	5	10	1	4	5	6	9	15
					e2	7	7	14	6	7	13	13	14	27
					Total	12	12	24	7	11	18	19	23	42
			d2	E	e1	6	7	13	7	6	13	13	13	26
					e2	7	5	12	3	3	6	10	8	18
					Total	13	12	25	10	9	19	23	21	44
			Total	E	e1	11	12	23	8	10	18	19	22	41
					e2	14	12	26	9	10	19	23	22	45
					Total	25	24	49	17	20	37	42	44	86
	Total	D	d1	E	e1	11	12	23	17	7	24	28	19	47
					e2	13	22	35	13	17	30	26	39	65
					Total	24	34	58	30	24	54	54	58	112
			d2	E	e1	13	13	26	9	14	23	22	27	49
					e2	16	13	29	13	12	25	29	25	54
					Total	29	26	55	22	26	48	51	52	103
			Total	E	e1	24	25	49	26	21	47	50	46	96
					e2	29	35	64	26	29	55	55	64	119
					Total	53	60	113	52	50	102	105	110	215

Analysis by Cell and Nest. Construction of the table, graph and semantic map.
Semantic analysis of the structure ($_2$A/$_2$B)

In this section we will see several examples of semantic analysis.

The following structure of table 48 ($_2$A/$_2$B)l has two nests and three TPs, we will see the construction of the semantic table for the analysis of the nests

Table 48. Structure ($_2$A/$_2$B)l

A Genre	a$_1$ Male	B marital status	b1 Single	N	53
			b2 Married	N	60
			Total	N	113
	a$_2$ Female	B marital status	b1 Single	N	52
			b2 Married	N	50
			Total	N	102
	Total	B marital status	b1 Single	N	105
			b2 Married	N	110
			Total	N	215

The following Box 13 shows the semantic table of structure H, the first column shows the frequency item f_2, the second the quantity N, the third the percentages obtained on the three TP of each semantic, placed in the fourth column, the remaining fifth and sixth columns correspond to the subject and predicate statistics, 46.9% of the men are single.

Box 13. Semantic table of the structure H= ($_2$A/$_2$B)l

Ítems	N	%*	TP	of	are
a$_1$b$_1$ Single Men	53	46,9	113	a$_1$ Men	b$_1$ Single
		50,5	105	b1 Single	a$_1$ Men
		24,7	215	TT	a$_1$b$_1$ Men Single
a$_1$b$_2$ Married men	60	53,1	113	a$_1$ Men	b$_2$ Married
		54,5	110	b$_2$ Married	a$_1$ Men
		27,9	215	TT	a$_1$b$_2$ Married Men
a$_2$b$_1$ Single Women	52	51,0	102	a$_2$ Women	b$_1$ Single
		49,5	105	b$_1$ Single	a$_2$ Mujer
		24,2	215	TT	a$_2$b$_1$ SingleWomen
a$_2$b$_2$ Married Women	50	49,0	102	a$_2$ Women	b$_2$ Married
		45,5	110	b$_2$ Married	a$_2$ Women
		23,3	215	TT	a$_2$b$_2$ Married Women

The semantic table allows exposing the statistical subject and predicating clearly without the chaos of percentages that are presented in conventional tables, it also allows the analysis by cell for small tables.

Given the large amount of information contained in the totals, we also construct relationships between them as ratios between them.

Box 14. Structure-to-total ratios H= $(_2A/_2B)l$

Total	N	Razones	
Men	113	1,1	Men to women ratio
Women	102		
Singles	105	0,95	Ratio of single to married
Marrieds	110		
Men	113	52,56	% of total
Women	105	48,84	
Total	215		

Semantic analysis of the structures T= [($_2A/_3G$)|$_2C$.]s and T'= [($_3G/_2A$)|$_2C$.]s redistributed.

The structures T in table 48 and T' in table 49 are equivalent as structures because their totals are maintained but redistributed by the overlapping effect of the nesting $_2A/_3G$ a $_3G/_2A$ which changes the strategy of the tabular analysis. In the first place the TPs are maintained $_3C_3+_3C_2+_3C_1= 1+3+3 = 7$ per item, in the second place the quantity of nests changes by the redistribution, in the structure, in T there are 4 that are formed because each trio of items of G closes with one of A, $a_1(g_1,g_2,g_3)$ and $a_2(g_1,g_2,g_3)$ crossed with the 2 of C, but in the structure T' of table 19 there are 6, $_3G$ by $_2C$; thirdly and as a consequence of this, the totals are redistributed as can be seen. So the analysis strategy changes according to the objectives.

Table 49. Structure T= [($_2A/_3G$)|$_2C$.]s

($_2A/_3G$)|$_2C$				C. Housing tenure		
				c_1 Has	c_2 Has Not	Total
				N	N	N
A. Genero	a_1. Male	G. Marital Status	g1. Single	21	18	39
			g2. Married	22	20	42
			g3. Divorced	21	11	32
			Total	64	49	113
	a_2. Female	G. Marital Status	g1. Single	22	9	31
			g2. Married	21	17	38
			g3. Divorced	22	11	33
			Total	65	37	102
	Total	G. Marital Status	g1. Single	43	27	70
			g2. Married	43	37	80
			g3. Divorced	43	22	65
			Total	129	86	215
		G. Marital Status				
		G. Marital Status				

We have already seen the analysis by cell or semantic, the analysis by nest is made with respect to the percentage common to the nest, in the structure T the first nest is analyzed with respect to the total *Male* (113 frequencies) that corresponds to layer a1, in the second structure T' the first nest is analyzed with respect to the total of *bachelors* (70 frequencies) of the major layer g1; the semantics of the first nest of T are man-single-(has-no housing), man-married-(has-no housing) and man-divorced-(has-no housing), those of structure T' are man-single-(has-no housing) and woman-single-(has-no housing). Note that the effect of the redistribution qualitatively changes the focus of the analysis, so that a semantic analysis strategy is the redistribution of the layers in the family $_2A_3G_2C$,, with its 6 classes (classes = 3!).

Table 50. Structure $T' = [(_2A/_3G) |_2C.]$s redistributed

			C. Housing tenure		
			c_1 Has	c_1 Has	Total
			N	N	N
G. Edo Civil	g_1 Single	A Genre	a_1 Maleu	21	18	39
			a_2 Female	22	9	31
			Total	43	27	70
	g_2 Married	A Genre	a_1 Male	22	20	42
			a_2 Female	21	17	38
			Total	43	37	80
	g_3 Divorced	A Genre	a_1 Male	21	11	32
			a_2 Female	22	11	33
			Total	43	22	65
	Total	A Genre	a_1 Maleu	64	49	113
			a_2 Female	65	37	102
			Total	129	86	215

The formation of classes is a methodological strategy to divide a tabular structure and make it closer to the analyst.

Let us now look at the semantic tables of both structures for the first nest of T and T'. The percentages are self-explanatory as they are accompanied by the statistical predicates.

Box 15. Semantic table of the first nest of the structure $T = [(_2A/_3G) |_2C.]$s

Semántic	N		TP	%	
	With housing	Homeless	Men	With housing	Homeless
Men solteros	21	18	113	18,58	15,93
Married men	22	20	113	19,47	17,70
Divorcied men	21	11	113	18,58	9,73

Box 16. Semantic table of the first nest of the structure $T' = [(_2A/_3G)|_2C.]s$ redistributed

Items	N		TP	%	
	With housing	Homeless	Single	With housing	Homeless
Single men	21	18	70	30,00	25,71
Single woman	22	9	70	31,43	12,86

Semantic analysis of the structure Σ= (₂A/₂B)|(₂C/₂D) and tabular segmentation.

In the following example the structure of table 51 is made a little more complex by adding the variable Degree of Instruction, it has the form $(_2A/_2B)|(_2C/_2D)l$.

First we calculate the TP = $_4C_1+_4C_2+_4C_3+_4C_4$ = 4+6+4+1 = 15, then we determine the number of nests which are 4 generated by the major layers $_2A$ in L position and $_2C$ in S position, each nest has 4 semantics for 15 TP would be 60 percentages per nest something very extensive if we do not have an analysis objective. The first nest is composed by the layers of men who have housing, the second by those who do not have housing, the third by women who have housing and the fourth by women who do not have housing. Of these nests, marital status and educational level are of interest; the strategy of redistributing the layers can be used if it is desired that the nests contain other information. The reader can do the analysis by nest or by any cell of interest by constructing semantic tables. In this section we will analyze the totals and segmentation of the structure.

Tabla 51 Structure $(_2A/_2B)|(_2C/_2D)$

					C. Housing tenure								
					c₁ Has			c₁ Has			Total		
					D level of education			D level of education			D level of education		
					d₁ Prof	d₂ Téc	Total	d₁ Prof	d₂ Téc	Total	d₁ Prof	d₂ Téc	Total
A Gender	a₁ Male	B.Status Marital	b₁ Married	N	13	15	28	11	14	25	24	29	53
			b₂ Single	N	13	23	36	12	12	24	25	35	60
			Total	N	26	38	64	23	26	49	49	64	113
	a₂ Female	B.Status Marital	b₁ Married	N	18	17	35	8	9	17	26	26	52
			b₂ Single	N	11	19	30	10	10	20	21	29	50
			Total	N	29	36	65	18	19	37	47	55	102
	Total	B.Status Marital	b₁ Married	N	31	32	63	19	23	42	50	55	105
			b₂ Single	N	24	42	66	22	22	44	46	64	110
			Total	N	55	74	129	41	45	86	96	119	215

Cuadro 17. Calculo de algunas razones entre totales de la estructura $(_2A/_2B)\,|\,(_2C/_2D)$

Totals	Reason	Interpretation
Men with housing / without housing	1,31	There are 1.31 males with housing for every 1 without housing.
Women with / without housing	1,76	There are 1.76 females with housing for every 1 without housing.
Single with / without housing	1,59	
Married with / without housing	1,16	
Divorced with / without housing	1,95	
Men to Women with housing	0,98	There are 0.98 men for every 1 woman with housing.
Men to women without housing	1,32	Interpretation

As can be seen in Table 17, the analysis of the totals can be extensive but can be expressed in semantic tables.

The structure $\Sigma= (_2A/_2B)\,|\,(_2C/_2D)$ can be segmented in several ways[12] , by partitioning or by tabular division, in this case there is no stacking of variables that allows partitioning by variables however the table can be split by the totals of the layers (Gender and Tenure) found at the end of table 20, these forms are[13] $\Sigma T_L= (_2Bt\,|\,(_2Ct/_2Dt)s)$ and $\Sigma T_S= (_2At/_2Bt\,|\,_2Dt)s)$, TL indicates that the totals are those in L position and TS that are in S position; the tabular division can be done by the Gender or Housing Tenure layers, we will do it by the former, the structure Σ is divided into two Σa_1 and Σa_2, the former takes the form $\Sigma a_1= a_1/(_2B\,|\,(_2C/_2D)s)$ y $\Sigma a_2= a_2/(_2B\,|\,(_2C/_2D)s)$ where a1 is the semantic Male and a2 Female. The TP $= {_3}C_1+{_3}C_2+{_3}C_3 = 3+3+1 = 7$ being for $b_1c_1d_1$; $\{b,\,c,\,d;\,bc,\,bd,\,cd;\,bcd\}$, now how is it that the structure Σ has 15 TPs and the segments 7 each? Let us recall that the TPs of Σ are given over a semantics of 4 terms while the segments over 3, another issue is that in the division by totals it must be taken into account that they are totals of totals. Tables 21, 22 and 23 below show the tabular segmentations by division and by partition of totals of Σ.

Table 52. Tabular Division $\Sigma a_1 = a_1/(_2B\,|\,(_2C/_2D)s)$

		C. Housing tenure								
		c_1 Has			c_1 Has			c_1 Has		
		D level of education			D level of education			D level of education		
		$d_1.$ Profess	$d_2.$ Technic	Total	$d_1.$ Profess	$d_2.$ technic	Total	$d_1.$ Profess	$d_2.$ technic	Total
		N	N	N	N	N	N	N	N	N
	$b_1.$Married	13	15	28	11	14	25	24	29	53
B Status Marital	$b_2.$ Single	13	23	36	12	12	24	25	35	60
	Total	26	38	64	23	26	49	49	64	113

---

[12] Procedure allowed by some conventional statistical programs
[13] Statistical programs can perform these procedures

Table 53. Tabular Division $\Sigma a_2 = a_2/(_2B\,|\,(_2C/_2D)s)$

		C. Housing tenure								
		c₁ Has			c₁ Has			c₁ Has		
		D level of education			D level of education			D level of education		
		d_1. Professional	d_2. Ttechnician	Total	d_1. Professional	d_2. technician	Total	d_1. Professional	d_2. technician	Total
		N	N	N	N	N	N	N	N	N
B Status Marital	b_1.Married	18	17	35	8	9	17	26	26	52
	b_2. Single	11	19	30	10	10	20	21	29	50
	Total	29	36	65	18	19	37	47	55	102

Table 54 Partitioning of the structure Σ by the totals in L, $\Sigma T_L = (_2Bt\,|\,(_2Ct/_2Dt)s)$

		C. Housing tenure								
		c₁ Has			c₁ Has			c₁ Has		
		D level of education			D level of education			D level of education		
		d_1. Profession	d_1. Profession	Total	d_1. Profession	d_1. Professional	Total	d_1. Professional	d_1. Professional	Total
		N	N	N	N	N	N	N	N	N
B Status Marital	b_1.Married	31	32	63	19	23	42	50	55	105
	b_2. Single	24	42	66	22	22	44	46	64	110
	Total	55	74	129	41	45	86	96	119	215

Semantic analysis of the structure $\Gamma=(_5A/_2B)|(_3C/_2D)$ and tabular segmentation.

The following structure Γ (see Table 55) is large for the sheet so we must segment it in order to analyze it. First of all the 2 items of B times the 5 of A give 10 rows, the 3 of C times the 2 of D give 6 columns of f_4 frequencies, that is 60 cells of f_4 divided into 15 nests ($_5A$ times $_3C$) on the other hand the ST subtotals are 5 of A plus 3 of B are 8 in the rows, 3 of C plus 3 of D are 6 in the columns for 8+6 = 14 subtotals, now from the generic semantic abcd we get TP= $_4C_1+_4C_2+_4C_3+_4C_4$ = 4+6+4+1 = 15 TP of cells for each semantic, remember that when we speak of ST we refer to the totals of the headings and the TP to the cells with values.

The variables that form the headings are A(Marital Status)=(a_1. single, a_2. unmarried, a_3. married, a_4. separated, a_5. widowed); B(Sex)=(b_1. male, b_2. female); C(Activity status)= (c_1. employed, c_2. unemployed, c_3. inactive) and D(Worked during the week)= (d_1. yes, d_2. no). If we proceed to the segmentation by some variable the forms would be as follows

1. Segmentation by major layers

   1.1. From Γ we extract A leaving the L-form intact, leaving $a_i/(_2B\,|\,(_3C/_2D))$, therefore the structure is segmented into 6 parts, single, unmarried, married, separated, widowed and total.

   1.2. The same happens if we extract C. The structure is $c_i/(_5A/_2B)\,|\,_2D))$ dividing Γ in 4 parts, employed, unemployed, inactive and total.

2. Segmentation by intermediate layers.

   2.1. If we segment by B (Sex), it is $b_i/(_5A\,|\,(_3C/_2D))$, segmented in 3 parts, males, females and total. We must take into account that as we have distributed the layers because now B remains as a major layer, from the initial structure $(_5A/_2B)\,|\,(_3C/_2D)$, the respective rows have been separated leaving A, that is, from graph 24 the male or female rows are eliminated depending on which one has become a major layer.

   2.2. The same happens if it is segmented by the nested D that passes to the major layer.

Tabla 55. Estructura $\Gamma=(_5A/_2B)\,|\,(_3C/_2D)s$

		C. ACTIVITY CONDITION											
		c₁. Employed			c₂. Unemployed			c₃. Inactive			Total		
		D. Worked during			D. Worked during			D. Worked during			D. Worked during the week		
		d₁. Yes	d₂. No	Total	d₁. Yes	d₂. No	Total	d₁. Yes	d₂. No	Total	d₁. Yes	d₂. No	Total
A Edo Civil	B Sex	N	N	N	N	N	N	N	N	N	N	N	N
a₁. Single	b₁. Male	177	3	180	2	44	46	0	717	717	179	764	943
	b₂. Female	178	5	183	0	42	42	1	820	821	179	867	1046
	Total	355	8	363	2	86	88	1	1537	1538	358	1631	1989
a₂. Joined	b₁. Male	103	1	104	0	11	11	0	3	3	103	15	118
	b₂. Female	44	2	46	0	9	9	1	68	69	45	79	124
	Total	147	3	150	0	20	20	1	71	72	148	94	242
a₃. Married	b₁. Male	459	7	466	0	35	35	0	117	117	459	159	618
	b₂. Female	219	5	224	0	10	10	1	403	404	220	418	638
	Total	678	12	690	0	45	45	1	520	521	679	577	1256
a₄. Separated or divorced	b₁. Male	29	0	29	0	3	3	0	9	9	29	12	41
	b₂. Female	51	1	52	0	11	11	0	36	36	51	48	99
	Total	80	1	81	0	14	14	0	45	45	80	60	140
a₅. Widowed	b₁. Male	7	0	7	0	2	2	0	12	12	7	14	21
	b₂. Female	21	1	22	0	2	2	1	147	148	22	150	172
	Total	28	1	29	0	4	4	1	159	160	29	164	193
Total	b₁. Male	775	11	786	2	95	97	0	858	858	777	964	1741
	b₂. Female	513	14	527	0	74	74	4	1474	1478	517	1562	2079
	Total	1288	25	1313	2	169	171	4	2332	2336	1294	2526	3820

In Box 18 below we have listed the nests of the structure in Table 24. They are 15 obtained from the amount of the 5 subtotals of the form $_5A$ by the 3 of $_3C$ these totals limit the nests. Recall that the difference between the subtotals and the TP is that the Firsts are totals lines while the Seconds are cells with totals that only contain the items of a given semantic, therefore the analysis by TP refers to the semantic with all its items and the analysis of the nests interprets the layers to which the semantic belongs.

Box 18. Nests of the structure $\Gamma=(_5A/_2B)\,|\,(_3C/_2D)s$

N°	Nest	Layers		Value
1	a_1c_1	Singles	Employed	363
2	a_2c_1	United	Employed	150
3	a_3c_1	Married	Employed	690
4	a_4c_1	Separated	Employed	81
5	a_5c_1	Widowed	Employed	29
6	a_1c_2	Single	Unemployed	88
7	a_2c_2	United	Unemployed	20
8	a_3c_2	Married	Unemployed	45
9	a_4c_2	Separated	Unemployed	14
10	a_5c_2	Widowed	Unemployed	4
11	a_1c_3	Single	Inactive	1538
12	a_2c_3	United	Inactive	72
13	a_3c_3	Married	Inactive	521
14	a_4c_3	Separated	Inactive	45
15	a_5c_3	Widowed	Inactive	159

All these nests have in common that they are people of both sexes and that they have worked or not the week of the interview, a quick view shows that nests 11, 3 and 13 of inactive single, busy-married and inactive-married are the most relevant therefore it can be deepened in the analysis of these nests to see which semantics characterize them.

We will now look at the TP analysis of the semantic a1b1c1d1 male-single-occupied-which-if-has-worked-in-the-week which has a high frequency in the nest.

SEMANTIC TABLE

Box 19 below shows the TPs for the semantic a1b1c1d1 male-single-occupied-who-has-worked-in-the-week. We particularly focus on the ones that have to do with a male layer (a1) and that we have colored in gray

N	the %	TP	Semántic	of	are	
177	98,3	180	$a_1b_1c_1$	Employed single male	d_1	if they have worked during the week
	49,9	355	$a_1c_1d_1$	Employed singles who have worked	b_1	males
	48,8	363	a_1c_1	Employed singles	b_1d_1	men who have worked during the week
	98,9	179	$a_1b_1d_1$	Single males who have worked	c_1	employed
	22,8	775	$c_1d_1b_1$	Employed males who have worked	a_1	singles
	22,5	786	b_1c_1	employed males	a_1d_1	singles who have worked
	18,8	943	a_1b_1	male singles	c_1d_1	employed who have worked
	8,9	1989	a_1	singles	$b_1c_1d_1$	employed men who have worked
	13,7	1288	c_1d_1	employed who have worked	a_1b_1	single males
	13,5	1313	c_1	employed	$a_1b_1d_1$	single males who have worked during the week
	49,4	358	a_1d_1	singles who have worked	b_1c_1	employed males
	22,8	777	b_1d_1	male who have worked during the week	a_1c_1	employed singles
	13,7	1294	d_1	who have worked during the week	$a_1b_1c_1$	single-occupied male
	10,2	1741	b_1	male	$a_1c_1d_1$	employed singles who have worked
	4,6	3820	TT	Total	$a_1b_1c_1d_1$	employed single men who have worked in the week

In the semantic table we read 98.3% of those who are single males and are employed, if they have worked during the week, this group represents 4.6% of the total, then follow other semantics related to single males and which are compared with or single females such as the Second which indicates that 49.9% of the employed single males who have worked are males. We thus see that the semantic chart relates the semantics associated with the objectives of an investigation. The semantic chart below visually shows the most important semantics.

In this section we show the segmentation of the structure Γ of table 55 that we alluded to earlier. We will only show them so that you can see the possibilities of Tabular Statistics when using TPs and equivalent tables.

Table 56. Structure Γ segmented according to occupational status. The employed
$\Gamma c_1 = c_1/((_5A/_2B)\,|\,_2D))$

| | | | | | D HAS WORKED DURING THE WEEK | | |
					d_1 Yes	d_2 Not	Total
A CIVIL STATUS	a_1 Single	B SEX	b_1 Male	N	177	3	180
			b_2 Female	N	178	5	183
			Total	N	355	8	363
	a_2 Unmarried	B SEX	b_1 Male	N	103	1	104
			b_2 Female	N	44	2	46
			Total	N	147	3	150
	a_3 Married	B SEX	b_1 Male	N	459	7	466
			b_2 Female	N	219	5	224
			Total	N	678	12	690
	a_4 Separated or divorced	B SEX	b_1 Male	N	29	0	29
			b_2 Female	N	51	1	52
			Total	N	80	1	81
	a_5 Widowed	B SEX	b_1 Male	N	7	0	7
			b_2 Female	N	21	1	22
			Total	N	28	1	29
	Total	B SEX	b_1 Male	N	775	11	786
			b_2 Female	N	513	14	527
			Total	N	1288	25	1313

Tabla 57. Structure Γ segmented according to occupation status. The unemployed
$\Gamma c_2 = c_2/((_5A/_2B) |_2D))$

| | | | | HAS WORKED DURING THE WEEK | | |
				d_1 Yes	d_2 No	Total
A CIVIL STATUS	a_1 Single	B SEX	b_1 Male N	2	44	46
			b_2 Female N	0	42	42
			Total N	2	86	88
	a_2 Unmarried	B SEX	b_1 Male N	0	11	11
			b_2 Female N	0	9	9
			Total N	0	20	20
	a_3 Married	B SEX	b_1 Male N	0	35	35
			b_2 Female N	0	10	10
			Total N	0	45	45
	a_4 Separated or divorced	B SEX	b_1 Male N	0	3	3
			b_2 Female N	0	11	11
			Total N	0	14	14
	a_5 Widowed	B SEX	b_1 Male N	0	2	2
			b_2 Female N	0	2	2
			Total N	0	4	4
	Total	B SEX	b_1 Male N	2	95	97
			b_2 Female N	0	74	74
			Total N	2	169	171

Tabla 58. . Structure Γ segmented according to occupation status. The inactive
$\Gamma c_3 = c_3/((_5A/_2B)|_2D))$

					HAS WORKED DURING THE WEEK		
					d_1 Yes	d_1 Yes	Total
A CIVIL STATUS	a_1 Single	B SEX	b_1 Male	N	0	717	717
			b_2 Female	N	1	820	821
			Total	N	1	1537	1538
	a_2 Unmarried	B SEX	b_1 Male	N	0	3	3
			b_2 Female	N	1	68	69
			Total	N	1	71	72
	a_3 Married	B SEX	b_1 Male	N	0	117	117
			b_2 Female	N	1	403	404
			Total	N	1	520	521
	a_4 Separated or divorced	B SEX	b_1 Male	N	0	9	9
			b_2 Female	N	0	36	36
			Total	N	0	45	45
	a_5 Widowed	B SEX	b_1 Male	N	0	12	12
			b_2 Female	N	1	147	148
			Total	N	1	159	160
	Total	B SEX	b_1 Male	N	0	858	858
			b_2 Female	N	4	1474	1478
			Total	N	4	2332	2336

Tabla 59. . Structure Γ segmented by occupational status. With respect to the total
$\Gamma_{Tg} = tg/((_5A/_2B)|_2D))$

				HAS WORKED DURING THE WEEK		
				$d_1.$Yes	$d_1.$Yes	$d_1.$Yes
A CIVIL STATUS	a_1 Single	B SEX	b_1 Male	N	179	764	943
			b_2 Female	N	179	867	1046
			Total	N	358	1631	1989
	a_2 Unmarried	B SEX	b_1 Male	N	103	15	118
			b_2 Female	N	45	79	124
			Total	N	148	94	242
	a_3 Married	B SEX	b_1 Male	N	459	159	618
			b_2 Female	N	220	418	638
			Total	N	679	577	1256
	a_4 Separated or divorced	B SEX	b_1 Male	N	29	12	41
			b_2 Female	N	51	48	99
			Total	N	80	60	140
	a_5 Widowed	B SEX	b_1 Male	N	7	14	21
			b_2 Female	N	22	150	172
			Total	N	29	164	193
	Total	B SEX	b_1 Male	N	777	964	1741
			b_2 Female	N	517	1562	2079
			Total	N	1294	2526	3820

Tabla 60. . Structure Γ segmented according to total marital status. With respect to the total
$\Gamma_{Tg} = tg/(_2B|_3C/_2D)$

		C Activity condition											
		c_1 Occupied			c_2 Unoccupied			c_3 Inactive			Total		
		D Worked during the week			D Worked during the week			D Worked during the week			D Worked during the week		
		d_1 Yes	d_2 No	Total	d_1 Yes	d_2 No	Total	d_1 Yes	d_2 No	Total	d_1 Yes	d_2 No	Total
		N	N	N	N	N	N	N	N	N	N	N	N
B Sex	b_1 Male	775	11	786	2	95	97	0	858	858	777	964	1741
	b_2 Female	513	14	527	0	74	74	4	1474	1478	517	1562	2079
	Total	1288	25	1313	2	169	171	4	2332	2336	1294	2526	3820

Methodological Aspects of the Tables

The traditional statistical methodology required that after the definition of the variables, their items, measurement mode and questionnaire, the Basic Tabulation Plan, that is, the planning of the data output; these outputs included a previous analysis of the variables and above all the form that the tables or configuration would have, since the processing was cumbersome and delicate in such a way that mistakes in the processing required machine time and programming by its operators.

This has changed radically today, the data analyst or researcher no longer needs (unless he does not master the statistical software not being justified today) another analyst to program the output of the results, likewise the researcher no longer needs to perform a previous Tabulation Plan, so that this methodological burden is transferred to the Exploratory Data Analysis (EDA) which is the analysis prior to that performed on the research..

Exploratory Data Analysis. EDA

This initial analysis was born of the computer revolution in the second half of the last century, its initiator was J. Tukey (1915-2000) who transformed the traditional descriptive analysis that was based on the reduction of data to class intervals and histograms, construction of ogives and in general Statistics for Grouped Data to the non-aggregated analysis of the data and therefore to the interrelation in an individual way. The classic problem was that a large volume of data to be analyzed had to be grouped because it was impossible to handle manually, but with the development of software and the power of hardware could be analyzed each data and its location in the sample giving rise to the AED with its techniques of analysis of stems and leaves, box plots, missing values among others that replaced the old ones and among other achievements the statistical tables acquired versatility, ease of construction, but at the same time complexity and need for a Tabular Statistics that allows both exploratory and final analysis of the study. The traditional tables as we have said refer their totals on the edges as fixed totals, reminiscent of the mode of analysis to build models of association and correlation. Tabular statistics makes the tables more dynamic by making use of semantics and totals and no longer displaying a meaningless cloud of numbers and a complexity of the tables that makes them not very usable. Statistical programs and spreadsheets allow the mobility of the structures being the basis for the tabular analysis we have seen.

Index of Terms

Referencias Bibliográficas

Casanova, H. (2013). *Estadística Tabular. Estructura, tipos y métodos para hacer tablas y gráficos. Método SL para hacer tablas.* Caracas, Venezuela: Escuela Venezolana de Planificación. Colección Notas Docentes N° 4.

Casanova, H. (2017). Método SL de construcción de tablas estadísticas como interase manual de la brecha digital en educación. *Docencia Universitaria, VXIII*(1 y 2), 73-93.

Casanova, H. D. (Mayo/ Agosto de 2022). Estadística y Ciencia de Datos ¿Qué hay de nuevo? *Actualidad Económica*, 55-75.

Cohen, M., & Nagel, E. (2000). *Introducción a la Lógica y al Método Científico* (Vol. 2). Buenos Aires, Argentina: Amorrortu.

Copy, I., & Cohen, C. (2013). *Introducción a la Lógica* (Segunda ed.). México: Limusa.

Díez, J. A., & Moulines, C. U. (1999). *Fundamentos de Filosofía de la Ciencia.* Barcelona, España: Ariel, S.A.

Hey, T., Tansley, S., & Tolle, K. (2009). *Fourth Paradigm. Data-Intensive Scientific Discovery.* Whashington, EEUU: Microsoft Corporation.

Taylor, J. R., & Kinnear, T. C. (1981). *Investigación de Mercados. Un enfoque aplicado.* Mexico, Mexico: Mc Graw-Hill.

Yong Varela, L. A. (enero-junio de 2004). Modelo de aceptación tecnológica (tam) para determinar los efectos de las dimensiones de cultura. *Revista Internacional de Ciencias Sociales y Humanidades, XIV*(1).